Luca Galofaro

MW00711676

Digital Eisenman

An Office of the Electronic Era

Birkhäuser – Publishers for Architecture
Basel • Boston • Berlin

Translation into English: Lucinda Byatt, Edinburgh

A CIP catalogue record for this book is available from the Library of Congress, Washington D.C., USA

Deutsche Bibliothek Cataloging-in-Publication Data

Digital Eisenman : an office of the electronic era / Luca Galofaro.
[Transl. into Engl.: Lucinda Byatt]. - Basel ; Boston ; Berlin : Birkhäuser, 1999
ISBN 3-7643-6094-1 (Basel ...)
ISBN 0-8176-6094-1 (Boston)

Original edition:
Eisenman digitale. Uno studio dell'era elettronica (Universale di Architettura 55, collana diretta da Bruno Zevi; La Rivoluzione Informatica, sezione a cura di Antonino Saggio).
© 1999 Testo & Immagine, Turin

© 1999 Birkhäuser – Publishers for Architecture, P.O. Box 133, CH-4010 Basel, Switzerland.
Printed on acid-free paper produced of chlorine-free pulp. TCF ∞
Printed in Italy
ISBN 3-7643-6094-1
ISNB 0-8176-6094-1

9 8 7 6 5 4 3 2 1

Contents

To Daria

I would like to thank Peter Eisenman for his encouragement and for having kindly provided the iconographic material; Richard Rosson for having answered all my questions about work in the office; Julliette Cesar for having helped me to assemble all the iconographic material; and all those who have shared this experience in Eisenman Architects: Diana Giambiagi, Jean-Cedric de Foy, Jan Henrik Hansen, Matthias Hollwich, Nikola Jarosch, Peter Lopez, Elisa Rosaria Orlanski, Georg Mahnke, Philipp Mohr, Maria Rita Perbellini, Sven Pfeiffer, Bernd Pflumm, Marco Pirone, Christian Pongratz, Ingeborg Rocker, Bettina Stolting, Martin Ulliana, Selim Vural.

My brother Marco has virtually prolonged my presence in the office and his help during post-production has been vital.

Lastly, my thanks to Antonino Saggio for having enabled me to record this experience.

This book could be described as a reportage; it is not the tale of a conscience, but rather its more immediate and simultaneous reportage.

PETER HANDKE

A starting point

The following paragraphs were not initially meant to appear in this form; I started to write to understand, or rather to follow a line of analysis, without even knowing that it would take so long. But then pictures, cuttings, readings and much else were added to my words day after day. A few months later, when I found myself leafing through this material, I realized that I could try to sum up all the pages by following a line of thought, which has guided me right to the end and has shown me Peter Eisenman's architecture day after day. My collection continues, I go on recording other events and other thoughts, all of which have influenced these pages until now. My research is guided by a particular line of inquiry: the attempt to explain how the computer can enter a designer's work and at the same time how it acts on the architectural space created by him. It might seem that anyone who is not born in the electronic era cannot handle the infinite possibilities of the computer and therefore cannot use it correctly. My personal experience in Eisenman's office demonstrates the opposite, or rather attempts to assess the slight architectural changes and major ideological implications of the digital instrument.

The structure of this work calls for a certain commitment by the reader, who can follow the leaps and changes of course by choosing an extremely personal interpretation, but I am sure that a certain degree of fragmentation will prove advantageous in the end. Eisenman's words, taken from interviews and theoretical works, are given special emphasis throughout the work and all the material is the outcome of an attempt to synthesize his design work and the exchanges of opinion that took place with project managers during the design process. Every bibliographical reference is particularly important. The use of images serves to underline the main themes; it helps to recall and reconstruct through the direct visualization of a method and in some cases even provides greater emphasis than the main text. It will not be possible to follow every single topic of discussion, and likewise it is certainly not possible to follow Peter Eisenman completely in his architecture and writings. We can only attempt to grasp, assimilate and subsequently metabolize an idea of modernity.

Critical architecture

Ongoing criticism

If we look at cinema, music, art in general, we need to understand, go beyond and combine our ideas with others, even if the latter belong to different ways of working. This is mosaic culture and one of the possible roads of post-modern man. Eisenman embodies this through his commitment in the field and undertakes continuous research aimed at re-establishing the image of space or, better still, translating it from the traces of a past that is still present and calls for critical interpretation. Each traditional component is rethought by Eisenman in a sort of constant discussion with himself and with contemporary architectural debate.

In his book *Real Presences*, George Steiner hypothesizes a utopian society where all discussion of works of art in general is prohibited. A society where there is no mediation between works of art and their users, where it would be impossible to read a review of a film or an article on the latest building. There would be no more comments, or worse still, comments on comments. We would be faced with "a republic of writers and readers where the meeting point between the creators and their audience would no longer be mediated by professional opinion makers" (Steiner 89).

Today you need only pick up a newspaper to replace direct experience with information on exhibitions and books; in Steiner's republic these pages of criticism would be reduced to mere lists: catalogues and guides. But what would art be like without the commentary? Steiner affirms that comments would still exist; in fact, they would be included directly in other works or in their execution. Playing a symphony also represents criticism of the symphony itself, but, in contrast to the music critic, the players invest their own being in the interpretative process; its interpretation becomes an act of responsibility because the player is called directly into question.

Any art form is an act of criticism because an artist constantly quotes and re-elaborates other people's material, drawing on different disciplines. This "ongoing criticism" is particularly evident in Peter Eisenman's projects. His undisputed ability to comment on the contemporary transforms his critical reflections into a unique form of architecture, characterized by constant development, and shows that "the best interpretation of art is art" (Steiner 92).

Peter Eisenman regards buildings as lessons in architecture, first Terragni, then Palladio, inheriting from each other. Terragni searches for the unconscious Palladio, the aspects that even Palladio himself could not understand in his own work. He searches for it, to destroy it and deny it, in the same way that Eisenman looks for Terragni by inventing him through the interaction of Eisenman-Terragni-Eisenman.

All this is essential to our understanding of how the office works and to our interpretation of Eisenman's architecture. Following the tracks that Eisenman gives us day by day is not merely important, it is a fundamental mental exercise.

During the seminars held inside his office, Eisenman once presented the plan of Campo Marzio (1762) as an example: this was justified by the fact that Piranesi had successfully separated the reality of the plan from the *sign* of that reality.

> Piranesi's drawing contains several different memories, the real memory of Roman ruins, the imaginary memory of Piranesi's imagination, the memory of buildings that still exist and the memory of the buildings moved by Piranesi (Eis 1).

We must assimilate his teaching, look at architecture with a critical eye and be transgressive not formally, but rather using an approach capable of diversifying space in physical terms.

Giovanni Battista Piranesi, The prisons: first stage.

Aronoff Center, interior.

> Produce an architecture that asks us to use our body again. Architecture as excess; as a possible condition beyond function, beyond iconography (Eis 2).

The critical interpretation of Piranesi's architecture becomes a founding element of his design method, in which interpretation and comment blend in a single creative act capable of reinventing the rules of his own work. According to Eisenman, architecture has always been a critical act, using two different methods. The first is the attempt to oppose current practice and accepted rules, trying to change the conditions imposed by space and time. The second is rooted in Kant's definition of criticism: namely, criticism "is the condition that makes existence possible." In the case of architecture, this definition transports architecture out of the sphere of the useful or the meaningful.

Contrary to current opinion, critical architecture is not the product of the *Zeitgeist* but rather the result of an opposite concept. Over the years, this type of architecture has been labelled a posteriori in a number of completely different ways, limiting its potential expression. But leaving aside all ideologies and all attempts at classification, the only common element in works by critical architects is the absolute conviction that form has some sort of positive influence on social reality. The collapse of the ideologies that guided part of the avantgarde appears to have forced architecture into a secondary role, global capital has replaced all ideology, extracting life-giving sap from architecture.

> From the end of modernism onwards, architecture has continued to waver between one stylistic extreme and another, between deconstructionist extravaganzas and day-to-day banality – in its search for an up-to-date form (Eis 3).

To Eisenman's way of thinking, an up-to-date form has always been linked to the meaning of architecture itself, a meaning that today seems to have been replaced by media instruments, hindering all symbolic and significant reasoning. Does architecture still have a meaning? What forms must it assume when it loses its iconic functions? Once again current trends set a limit to the development of critical architecture. On the one hand, there are those who think that architecture should essentially deal with technological infrastructure capable of being a medium in its own right, thus relegating it to a pure technical extension of capital.

> The possibility of a critical meaning is completely eliminated, because there is only one solution to the problem of infrastructure: the need for precision rules out the space for interpretation (Eis 4).

Those who believe in the critical role of architecture, on the other hand, wholeheartedly affirm that form is still a vehicle for meaning. For Eisenman, architecture must re-launch its own form, and to do so we must experiment with a post-critical architecture. In the exhibition entitled *Formless: a User's Guide*, Yves-Alain Bois and Rosalind Krauss suggest that "the point is the possibility of preserving the individuality of objects when they lose all their conventional means of legitimisation" (Krauss 96).

> Architecture must be capable of questioning both the traditional way of expressing meaning and of solving the problem of function. And thus both the social function and its current modes of legitimisation (Eis 5).

Eisenman as Piranesi

Post-critical architecture is Eisenman's attempt to lay the foundations for a search that he hopes will satisfy his desire to rewrite the canons of a discipline.

Given that Piranesian architecture has always represented a critical act, Eisenman appears to use it not only as a visual similarity, but also a starting point, almost a figurative analogy based on Piranesi's method of conceiving his architecture. Piranesi as a critical architect, Eisenman as a post-critical architect. Piranesi using perspective, Eisenman using the computer, mock-ups and 3D models, both in search of an

Aronoff Center, interior.

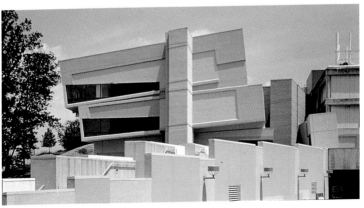

Aronoff Center, exterior.

14

appropriate instrument to control their taste for superimposition and the reinterpretation of memory, the endless assembly of multiple vanishing points and the breakdown of the
means of expression and lines into an infinity of moving seg-
ments. Both are searching for the compactness of dissolving
forms. All these aspects can be seen in both the *Carceri* and
Eisenman's buildings. This process is even more artful in the
Columbus Center, whereas the use of the computer turns the
Aronoff Center into a fluid system in movement. Eisenman
models space and doubles the image of objects, which dissolve like the physical elements of representation.

12-13

> Stone is moved from stone, but has preserved the figurative form of
> stone. The overlapping vanishing points border on the madness of
> narcosis, but every ring of these perspectives, which together make
> one's head spin, is in itself quite naturalistic.

These are the words used by Sergej Ejzenštejn to describe
Piranesi's vision and they also anticipate a journey into
Eisenman's architecture. Juxtaposed individual elements lose
their shape and are transformed into independent blocks of
pale and dark colors, rather than light and shade, which create
a new world of space, volumes and their intersections. Color is
used to reinforce the contrast and transition between reference systems and to dematerialise spatial tension. The assembly technique is applied to the instruments used to create these
buildings: first diagrams and then CAD modelling attempt to
move harmoniously from some forms to others, and the
imprint they leave on each other underlines their emotive par-
ticipation and involves us in the space.
Piranesi makes unique use of space, his visions take shape
through the interaction of the same architectonic motive. The
dislocation of the subject occurs through the spatial manipulation of individual elements that, repeated to infinity, compel
the onlooker to perform an act of perception that embraces
the entire design in an attempt to reconstruct his own itinerary
and his own position in virtual terms. At this point, space
guides us, involves us and transports us to a world made up of

interlinked and intersecting floors, and when the floors are pushed forward, out of the drawing, we feel threatened and therefore again in direct relation with space. Perspective is not only used as an instrument, but an interrupted system of spatial interpretation; in the *Carceri* it is difficult to find a perspective view that moves into depth without interrupting the spatial continuity. Likewise, in Eisenman's architecture, the pillars, stairs and volumes break all perspective movement. In the Aronoff Center the fluidity of the central passage is interrupted by the columns rising up in the center of a staircase, and the volumes that intersect the passages; we are compelled to move past them, following the winding walls: the space becomes a constant discovery. Beyond the pillar or staircase, the movement starts again until it meets a new obstacle, creating a space whose proportions are constantly changing.

But what is there about Eisenman's method that is reminiscent of Piranesi, and how does the American architect use these rules in his architecture? Piranesi used perspective to model transgressive buildings capable of overturning established rules. In Eisenman, the means change but the end is the same: perspective is no longer the guide used to create architecture. The new technologies hold the key to this change, experimentation becomes more complex and finds a major ally in the computer. Eisenman is not certain of the structure, but he guesses its potential and manages to follow it in a variety of ways.

Three-dimensional modelling means that you do not need to work by successive frames, but only on the space, which is sectioned later and then developed from the inside. While Piranesi uses perspective to place himself inside the space, Eisenman uses the computer and models to work simultaneously from inside and outside, constantly overturning accepted practice and vision. The computer enables him to work by following parallel but reciprocally influential routes, both characterized by movement. While Piranesi manages to recreate movement using a static image, Eisenman finds it by using a dynamic system. In Piranesi, it is our eye that guides our vision "and the final collision between this presumed trajectory

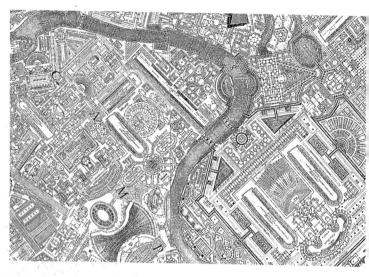

Above: View of Campo Marzio, G.B. Piranesi; below and facing page: urban competition for Klingelhofer Triangle, Berlin 1995. The project design stems from the relationship between different figures, in this case the mechanism of a clock and a micro chip changed by morphing. No figure dominates the other and these diagrams reveal a new intervention strategy for the city.

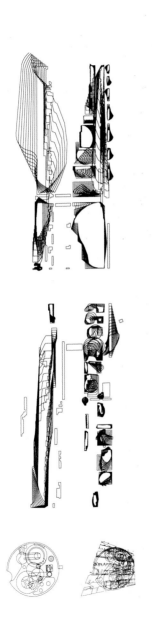

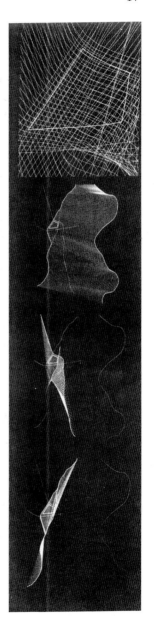

and the alternative trajectory produces the effect of thrust" (Tafuri). In Eisenman's architecture, even in his unrealized designs, it is the body that measures and feels space. The system of perception is structured in a complex way and all the senses intervene for the total control of space. Both are looking for something in architecture. Piranesi wholeheartedly aspires to draw a three-dimensional body from the flat surface of the mold. Eisenman tries to retrieve a two-dimensional, or better, an immaterial image from a three-dimensional reality. What they have in common is a breakdown of figurative and spatial continuity through their research focused on tradition. Many regard movement as one of the architectural themes of the twentieth century. Cubists and Futurists have tried to capture and stop it: Duchamp and Boccioni are two examples, but both use a series of individual images that are recomposed and overlap within the same space simultaneously, generating a temporal palimpsest.

Eisenman does not try to stop this movement, but to bring it up-to-date through a system of virtual representation. This manipulation of the forms is achieved through space and not through an inert material. In his creative process, the architectural image is only a frame for spatial complexity, it is just another way of destabilizing and underlining complexity as a system of expression. Piranesi's greatness as an architect does not rely on arches and stairs, but rather on his spatial inventions seen as exceptions to the rule. Another figurative and conceptual analogy is used by Piranesi in Campo Marzio: here we see the triumph of the method of arbitrary associations and the exclusion of any definite structure through the principles of aggregation.

"The recognition of some alignments only serves to highlight with greater clarity the 'triumph of the fragment' that dominates the shapeless piling up of sham bodies" (Tafuri). Moreover, Tafuri affirms that Piranesi's intention is to detect the birth of a meaningless architecture, cut off from all symbolism and every value outside architecture itself.

Eisenman also searches for a loss of meaning that will restore to architecture a real value, lost in contemporary society.

Eisenman sees Campo Marzio as an example of a new type of urban form that is defined through the relationship between figures. This type of urbanism may be regarded as the first modern plan because of its complex internal logic. It was the first town plan to emphasize an anti-hierarchical ideology. A conception that breaks the linear continuity with tradition from an extensive point of view, the space-time continuum where every step is directly linked to the one before it. The translation of this system into urban forms represents the idea of a relationship between figure and ground, where the solid and the void exist in a sort of dialectic tension.

> The development of thought has meant that the ideal of an intensive spatial relation now replaces the extensive idea of the evolutionary continuum (Eis 6).

Turning to today's situation, we can affirm that we now look at history in a similar light to Piranesi. The technical obsession with assembly represents an extreme attempt to undermine language and restore inventive freedom. Piranesi's transgression, his anticlassical interpretation of Roman monuments, is the same ongoing criticism adopted by Eisenman.

> Piranesi shows something different. His views of ancient Rome often distort the real dimensions of the buildings: a typical example is the view of the Pantheon square in which the Imperial rotunda is reduced, whereas a giant obelisk standing in the centre of the fountain towers over it. Piranesi's engraving illustrates a truth than goes further than reality. The Pantheon blends into the urban continuum, it mixes with the city (Tafuri).

Eisenman proposes an urban system based on the Piranesian concept of relations between figures in his project for the Klingelhofer Triangle in Berlin (Eis 7), but contrary to [17] Piranesi's invention for Campo Marzio, his system is not based on the reinterpretation of classic Roman forms; it is inspired by the development of contemporary thought, and in particular the profound transformation of forms that took

Both pages: Eisenman's office and the cover of an issue of Architecture New York.

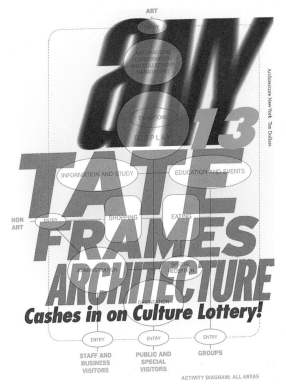

ART

ENTRY

ART HANDLING
CONSERVATION
AND COLLECTIONS
MANAGEMENT

EXHIBITIONS

DISPLAY

INFORMATION AND STUDY   EDUCATION AND EVENTS

NON
ART   ENTRY   SHOPPING   EATING

ADMINISTRATION   SPECIAL
RECEPTION

ORIENTATION

ENTRY   ENTRY   ENTRY

STAFF AND
BUSINESS
VISITORS   PUBLIC AND
SPECIAL
VISITORS   GROUPS

ACTIVITY DIAGRAM: ALL AREAS

Architecture New York Ten Dollars

TATE
FRAMES
ARCHITECTURE
Cashes in on Culture Lottery!

13

place between the mechanical era and the information era. As a symbol of this transformation, he chooses two analogous figures, each of which represents a different era: a clock mechanism and the digital microchip of a computer. Both represent a transition and are images of their time.

Piranesi uses Roman architecture, whereas Eisenman adopts a technique called *morphing* used in contemporary cinema: a transformation technique, a system capable of changing the chosen figures so that none is dominant; there is no relationship between the figure and the ground, but only between the overlapping figures, transformed into a new system of urban living. Architecture is reborn outside any recognized dogmas.

At this point, the project embarks on the development process using the digital system; the diagram is generated and influences the setting according to two scales of intervention. The first is at the city level, where morphing restores the site's position by creating a new relationship with the power centers. This leads to a new relationship with the government buildings at Spreebogen, the President's new residence, and the general HQ of Potsdamer Platz.

On the second scale, the building block, the process suggests a new way of living. The perimeter of the block was a valid urban concept when everyone living in the block shared the semi-public ground. But now this kind of space cannot be divided any longer owing to today's high urban densities, and the available space is constantly shrinking. Eisenman's project creates a new system of relations that are no longer tied to the relations between the residential block and public ground. It restores open spaces to everyone by making different use of the ground, optimizing its enjoyment.

This figure/figure system leads to the creation of a wide variety of building types, easily adapted to changing living requirements. An urban conception that is no longer deterministic, but adaptive, using the landscape as an active element capable of organizing its own layout; buildings are no longer seen as functional containers, but integrated systems capable of changing the very nature of the city itself. Total flexibility and a method that acts directly on the height, transparency, density and program, rather than on form. As in Piranesi's theoretical work and projects, Eisenman asks himself about the architect's role and his capacity to shape space. The logic of decomposition, much discussed and reinterpreted in the context of different works, reveals the discovery of the contradiction principle. The art of dialectic development, an architecture that evolves around itself and manages to renew itself by destroying each goal once it has been achieved. Eisenman's works give movement to Piranesi's views, forcing the language where necessary and searching in formal excess for a critical but not necessarily linear system of evolution.

How Eisenman works

41 West, 25th Street, New York City

Entering the office gives you a strange feeling. There are no filters or corridors; you immediately find yourself in a chaotic space, without partitions. You get the same feeling when you start working: no time is wasted on preliminaries, you dive straight into a project and everyone makes their own contribution. The moment of feeling out of one's depth, typical of a working situation of this kind, is replaced by an immediate emotional involvement in the design process. Eisenman loves football and, like a good coach, he encourages team play in which everyone does what they are best at. The design team rotates around a project manager, but all teams are supervised by Eisenman himself. In the large open-space office, everyone works together. It does not feel like a real architect's office, but almost like a university lecture room. This is what Eisenman wants because he sees himself primarily as a teacher. He has taught at various American universities, and still lectures at the Cooper Union, as well as on a daily basis in his workshop through the "didactic energy" of his architecture. His own buildings, but also traditional architecture, become textbooks; through their interpretation, architects develop a sense of criticism and an analytical dialectic that is vital to their work.

A CHURCH FOR THE YEAR 2000

The models help to see in advance, to follow the evolution of the building to be constructed. During the design phase, the models, diagrams and computer models intercommunicate. Each is inserted in a specific phase and affects the other, enriching it with new perspectives, and figurative and conceptual possibilities. The diagrammatic models become a theoretical reflection on the project and give it form.

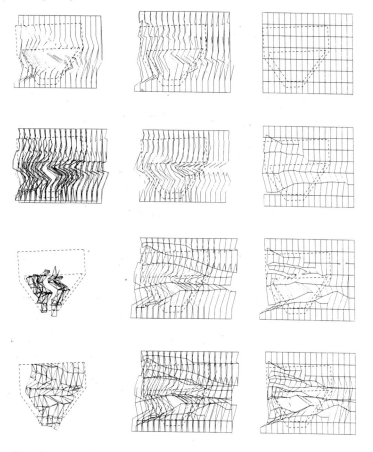

Development of diagrammatic models and parallel evolution of study models (facing page).

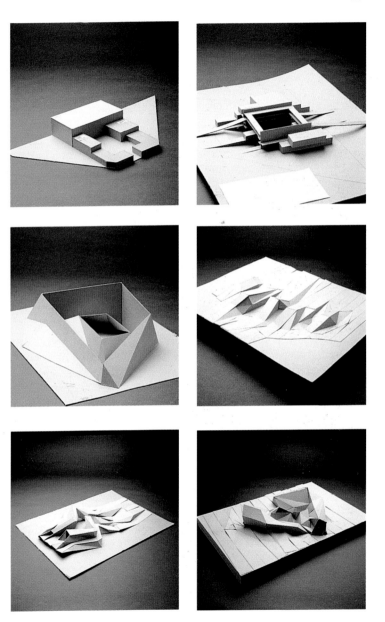

> The more I explain, the more I see. I believe that the task of explana-
> tion enables us to see more, to increase our capacity to acquire expe-
> rience, rather than to proclaim the truth [...] By forcing myself to
> explain, I learn more (Eis 8).

He is a teacher who transforms his theories and even his teaching method in a design process in which the presence of collaborators serves as an additional stimulus. Research proceeds in parallel with the formal elaboration of projects, each theme is developed with regard to its immediate architectural application and emphasis is placed on the correlated action of practice and theory. During the summer period, Eisenman holds seminars which highlight the elements to be developed during the year while the project is elaborated. Once a week, seated around the large meeting table, a group of different people, artists, critics, former collaborators, come to listen to his lessons, convinced of the importance of this ritual. This is just the start of a process that is not completed during the lessons, but continues with the awareness that thought crosses this working space even beyond architecture.

Almost as if to verify the entire process, the office produces a journal that broadens the debate to a international network of exchange, in which the individual national boundaries are lost in a global information system. *Architecture New York* is the best synthesis of events in the world of contemporary thought: architects, philosophers, writers, artists, a group with a variable composition (Lotus 80), that is not based on an ideology, but moves in the world of the end of ideologies. *ANY* publishes discussions on architecture and contemporary society in a process of cultural cosmopolitanism that, as Sanford Kwinter affirms, is now typical of many of the humanities. "It is only when other literary and philosophical disciplines project their critical theories onto architecture, using architecture as a source or reference point, that the latter becomes the central theme of critical thought" (Davidson 97).

The method through the project
One need only look around to understand the importance of

the three-dimensional control of space: models invade the whole office and the only drawings are the digital elaborations of the complex diagrams used to guide the modelling process. At first sight, this architecture might appear to be purely formal research, but everyday practice proves otherwise. The architectural project becomes the logical conclusion of a long speculative process that leaves tracks and opens up other feasible paths. In the same way as a hypertext, Eisenman's design method overlaps with an extremely consistent research procedure in which it is important to understand the value of architecture in relation to its theoretical and critical development. Reflection on modernity leads to the reinvention of a system in which each instrument plays a part in production and where information becomes a system capable of generating possible spaces.

When I joined the office, they were working simultaneously on completing a project for a church in Rome and starting designs for the BFL head offices in Bangalore, India, and the Jewish Museum in San Francisco. Collaborators are expected to look for possible meanings in the texts, writings and projects, but above all to identify the path leading to what Eisenman cannot yet see. This calls for a careful work of "translation", in which the written text describes what is drawn and the latter reflects what is written, and both always refer to one project. The work is one of constant dislocation towards an apparently non-sense determined by the constant and unforeseeable shift of direction, floors and spaces. This is the design process used to interpret the object to be constructed.

The next stage is to study the site using models of the area. During this preparatory phase, the bits are ordered to represent only what exists already, but they are not yet used in their infinite creative combinations. Modelling is used as both a creative strategy and a cognitive strategy. It is certainly not regarded as merely a mechanical backup operation, but invested with a complexity that can be compared with the sphere of the philosophy of science. Three types of models are used at every stage of the project: plastic models, diagrammatic models and computer models.

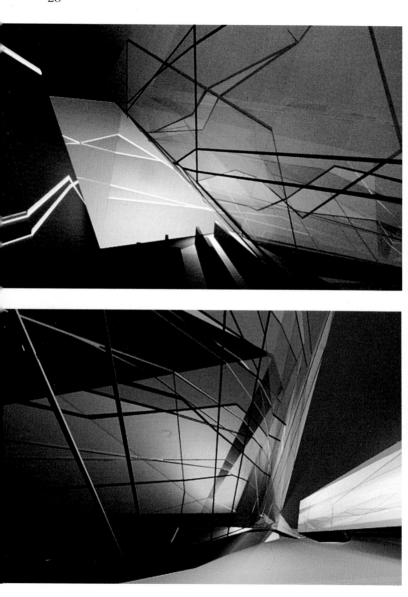

Computer models on a church in Rome, competition by invitation.

Computer models on a church in Rome, competition by invitation.

Plastic models

These are used to visualise all the formal hypotheses and they constitute the office's design system. The models help to see in advance, to follow the evolution of the building to be constructed; in this way, it is possible to monitor the development of a project by controlling every alteration in real time. This marks the start of the creative relationship between the cognitive-perceptive act and the figurative-operative intervention. In practice, Eisenman does not draw any distinction, except in terms of scale, between the model and the built object. He sees the model as an object deprived of the need to be lived in. Every phase starts with this instrument of representation, which interacts with other models. The possibilities of understanding a particular space are expounded beyond belief and relations with the client are also immediate. For example, the plastic model of a building can be used to simulate a view of the interior by a visitor, either standing still or moving, or it can simulate what we might call an endoscopic view of the building, with the help of digital photography or microcinematography.

Diagrammatic models

Non-iconic models are also used in the office, as well as iconic ones. There are various kinds of diagrammatic models: architectonic, philosophical-scientific and mathematical. Using the first it is possible to analyze aspects regarding the location of functions in a building or the vertical and horizontal connections between them; structure and function are important in this series of models, but not shape. Other models imported from other disciplines are not necessarily iconic, but represent the formal structure of the project.

> Instead of starting exclusively with a functional or typological diagram, we start from other types of diagram, like those using liquid crystals (church) or showing brainwave functions (Geneva library) (Eis 9).

In simpler terms, after an initial phase of study in which the project is developed using traditional methods, and different typological systems are examined, the latter are then over-lapped with deformation diagrams.

From a philosophical point of view, Eisenman supports the theories of several philosophers. He is attracted by non-Euclidean geometry, in particular Boolean geometry, used as the starting point for one of his most fascinating projects: the Carnegie Mellon Research Laboratories. He studies fractals, chaos theory, catastrophe theory, DNA, Leibnizian atoms, the behavior of liquid crystals. This knowledge and their principles guide the deformation and development of spaces. In the project for a church in Rome, the underlying scheme was that of two parallel bars (architectonic type of diagrammatic model) and the void compressed between them. The project developed from here. The church is based on two parallel premises: the first is the relationship between proximity and distance implicit in the concept of pilgrimage and in the idea of communication media; the second is the new relationship between man, God and nature. In fact, it is a natural form that is chosen to symbolize this condition, both of proximity and distance: the condition of liquid crystals, their suspension between the static crystal and a liquid state. The diagrams on liquid crystals offer the possibility of producing deformations capable of modelling the space. 24-29

The condition of "in-between," which is typical of the liquid crystal, also provides an opportunity for bonding with the site in a less arbitrary fashion. The architect already has an idea of the shape, but his research proceeds towards a so-called scientific definition. Once the reference has been chosen, the diagrams on the behaviour of liquid crystals overlap the typological diagrams and the building is born through their reciprocal influence and elaboration.

The diagrams show how the building emerges from the ground, taking shape according to the order of molecules in a crystal. The most interesting aspect are the possibilities offered by these deformations in a practical spatial application (as Eisenman has already shown in the Cincinnati building), where multiple layers and superimpositions are used to mold the space.

32

The diagrammatic structure follows the operations of human neurological activity, producing the architectural conditions capable of modelling space. The diagrams of cerebral functions are then superimposed on the site grid. Eisenman records the different elements and transforms them into design constraints, incorporating them in the mechanical process that generates the object directly from the site. The new building emerges between the landscape and objects, putting both conditions out of the focus in a single heterogeneous space.

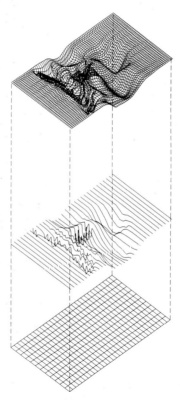

Conceptual diagrams, unification of the diagrams on memory: 1. Activity below the synaptic threshold. 2. Synaptic activity. 3. Consolidation of memory.

The process of synaptic activity is an example of a chaotic geometry that cannot be represented using the traditional order of Euclidean geometry. It is a heterogeneous and self-organized system, dominated by arbitrary combinations. Through repetition, synaptic activity produces extraordinary instances of the evolution process.

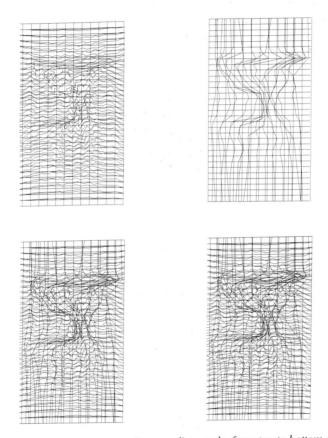

Conceptual diagrams, stages of intermediate study, from top to bottom and from left to right: 1. Superimposition of activity separation frequencies. 2. Superimposition of the tracings of memory in an emotional state. 3. Superimposition of tracings and frequencies. 4. Superimposition of diagrams on the site.

Once the relations between the tracings and the frequency have been established, they are transformed into a 3D relationship of solids and voids.

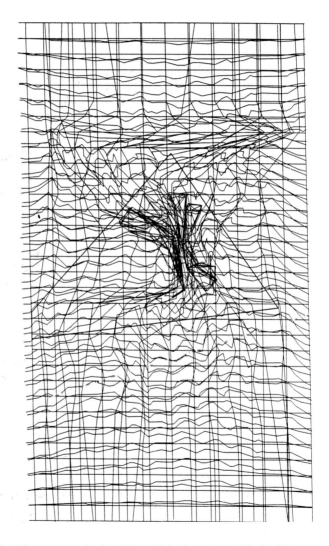

Final diagram: cerebral activity and the formation of the building.

The form of the building emerges from the successive superimposition of solids and voids, creating a complex relationship of interstitial spaces, to which the library functions are then assigned.

Study model and computer models of the building.

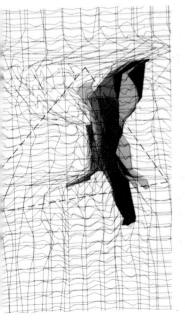

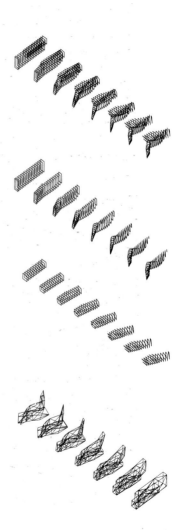

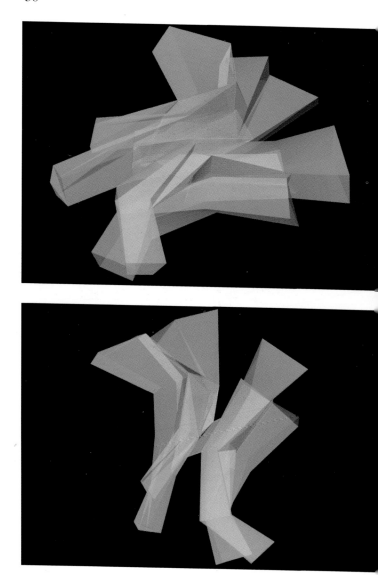

Models of the building body: the development process of the project entails constant passages between the computer and the designer who alters the models.

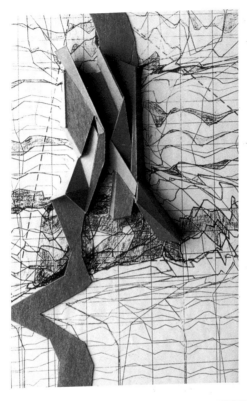

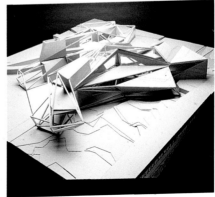

Model of the building body and computer models exploring the possibilities of the spaces between.

The shape of the church evolves from the ground, from a tangible reality, towards the sky and the infinite [...] it becomes the mediation between God and nature, between the physical and the infinite (Eis 9).

32-37 In the Geneva Library, on the other hand, he searches for formal justification in the diagrammatic structure based on the operations of the human memory, producing new architectonic conditions. The diagrams of neurological activity are superimposed on the general scheme and dictate the conditions capable of re-designing the library space. A study of pure form that arises through arbitrary reflection, but as always fathoms the themes of space between the unexplored interstices of architecture.

Eisenman discovered the potential of the void in the church in Rome and he continued his endless research in Geneva. If the solid is compressed and deformed, the void is also modified; it reacts by deforming itself and opposes the reactive force: the interstial spaces are generated by the compression of the void by the solid and the void by the void. This generation of space stems from the recording of lines and vectors. Diagrams like those recording the energy generated by the mass and the density of objects are used as parameters. The diagram evolves and becomes a vectorial system, a real computerized model.

Computerized models

These models from the electronic age are used with exceptional creativity in the office. They are much more pliable than traditional models in that they allow a more dynamic and immediate interaction. The computerized models can cover all the separate phases of a project and they bring together the different types of models used to date: graphics, dynamic diagrams and 3D views. This has resulted in the creativity of the design process, which is not solely based on computer equipment, but dialogues with all the different forms of representation used in order to develop an idea of architecture, seen above all as a place of invention and writing, rather than representation, in a new expressive form in which space is narration. The real goal of Eisenman's models is not to establish a universal theory, but to produce a critical text.

At this point, we may rightly wonder about the future of iconic and non-iconic modelling after the computer revolution: where will the boundary between the communication and design function lie, and how important will this interaction be in defining the new parameters?

We do not yet know, or we do not know enough about how this design creativeness will develop in the context of this new universe of modelling.

Genesis of a project

The work proceeds in a traditional pattern: the models overlap one another in a dialogue created through man's interaction with the machine. A strategy is chosen, an approach that is consistent with a series of different parameters, the place for the project is identified together with the theme to be developed and local tradition. For example, in India *mandala* is a sort of religious-aesthetic form very similar to Japanese *shui* used to select the orientation of the building and to identify a series of rules dictated by the forces acting on the place itself, depending on a principle linked to tradition and the energies present.

The first models are constructed using traditional layouts; the system of bars and rectangular volumes explores the possible forms of aggregation, searching for a size directly linked to the functional program. The main unit of volume is formed by the combination of the different basic functional units that guarantee the sequence of activities. The computer acts on diagrams, and the operator decides its development and genesis. The diagrams are interpreted, modified and reprocessed, metabolized in a precise direction by whoever is working on the project, using the original idea as a source of inspiration and guidance. Through this series of tracings, work starts on the production of 3D models that can be used to ascertain through comparison whether the design matches that of earlier models. In some cases, the photos of the chosen models are scanned into the computer and used as a basis for the two-dimensional drawings of the building. The computer comes back into play and is used to control forms and carry out further verification through the 3D drawings.

> It took us about two years to build up a good level of expertise with the machines used in the office (Eis 10).

At this point, the dialogue between operator, machine and model laboratory intensifies, there is a constant harking back to solutions rejected earlier that are transformed into new approaches to be followed. The relationship between hand drawn designs and the electronic management of the idea continues without a break and this dialogue leads to the birth of the first project. Every action is carefully verified using functional schemes and the models from the first stage.

The project appears in the distance, the real and virtual versions are constantly compared and the boundary lines dissolve in the process of realization.

How can we describe the virtual aspect of architecture and how can it be explored? It would certainly be simplistic to regard it as a mock-up of reality or a simulation of the physical aspects of architecture. Eisenman's office has set itself the goal of overcoming these definitions in favor of a "reinterpretation of the virtual as a present that is still missing in the physical environment" (Rocker 97).

The virtual is not a mere representation of something predefined, but suggests the notion of form as one of the many possible implementations of becoming virtual. "Its implementation is the creation-invention of a form starting from a dynamic configuration of forces and aims" (Levy 97), a real development of ideas that fosters the virtual itself.

A challenge, constant research into the value of shape, a developing process that suggests the possibility of the present through the past and the future. *Virtual* contrasts with *actual* and is contrary to *possible*, an evolving entity, an object understood as an elaboration of ideas and forms. Understood in this way, virtual is not the opposite of real, but the possibility, inherent in creation, of prefiguring what does not yet exist a moment before its physical realization. The virtual should therefore be seen as the series of problems accompanying a situation. For example, a tree is virtually present in a seed, and its transformation requires a process: its implementation. "The seed's problem is to grow a tree, the seed 'is' this problem, even if it goes further than this […] based on its inherent constraints, it must invent it, produce it depending on the circumstances encountered" (Levy 97).

Whereas reality resembles or, better still, reflects the possible, the actual responds to the virtual. Once we have understood the difference between realization (the occurrence of the possible) and implementation (the invention of a solution), we need only attempt to define virtualization understood as the dynamics and not as a way of being. Moreover, the notion of virtual is used, yet again, by Eisenman to destabilize the general notion of Cartesian space, where each object has a clear and precise form. In doing so, it overlaps with the concept of unform.

> The implementation of the virtual and its formal definition are independent of geometric limitations and of architectural proportions and constraints (Eis 10).

During the project phases, we constantly refer back to the theoretical problems discussed earlier, on which the office has been working for some time. From this, it can be understood how the American architect's thought is constantly evolving:

> I do not know a priori what some things are. The question of interstitiality has been analysed by many, I only try to understand its influence on the site and the programme (Eis 11).

Eisenman tries to understand, to elaborate a theory; during the design process, every project and every reflection, the preparation of every critical text adds something new to the experience. The models used come from other disciplines because architecture cannot provide suitable models to describe the complexity of the world. Until now, architectural forms have been bound to Euclidean geometry and Cartesian grids. This attempt to destabilize the status quo stems from the desire to understand and at the same time to foster a renewed importance of the object. Over the course of history, the most prestigious buildings have always been created in a contextual dialogue; Eisenman's work is aimed at a new dialectic, turning electronic culture into a method. The use of computers is not as important as the methods used by this culture, in which superimposition, collage and assembly are the instruments and systems used to reach the limit to be exceeded. In this sense Eisenman clarifies his position by just trying to comment, designing a creation that need not be emulated but understood, assimilated and reinterpreted. The system of the comments about comments will create a new world.

The organization of the working group for the offices of BFL Software Limited, Bangalore, India

The project in India will tackle a theme that still has no answer, even though it has already been analysed by this office: the question of interstitial space (Eis 12).

The "in-between" space has also been one of Eisenman's key themes of analysis and research. In the Wexner Center and Aronoff Center, instead of designing the building on a free site, he built by filling and reusing the void between the ground and the existing buildings. Eisenman always tries to make his own ideas evolve and therefore looks for the in-between space in all the possible combinations, even inside his own buildings, between the solids and the voids, giving them a different, or rather a more precise definition every time.

At the end of every project, having identified the problems Eisenman attacks them again in order to resolve them, even at the cost of reviewing a number of fixed points. Every end corresponds to a new beginning and a new form, and there is a deep-rooted continuity between every project, even if disguised by apparent discontinuity. The Wexner Center contains problems that will be resolved later, like the old idea of the frame and the non-existent conformation of the outside space. The solutions to this problem were found in the project for the church in Rome where the space between becomes a public space created through the manipulation of the building itself. At the first meeting, Eisenman organizes his collaborators and asks for their total commitment, underlining the importance of group work: all kinds of intervention are useful for the definition of the project. We must therefore play an active role in the entire process, and we are not here merely to execute orders. Eisenman does not know the ends to which his research is leading, or the shape of the final project. His initial suggestions and the themes to be analyzed are merely a starting point that may lead to different paths, thanks to the input made by his collaborators.

The client, the owner of a software company, is trying to relaunch his company by renewing its image, among other things; the functional program is very rigid and must be developed by the office. Planning in a country whose native traditions are completely different to those in the West is not easy. The first step, therefore, is to embark on intense research in order to build up the necessary knowledge regarding the meaning of architecture in Indian culture and its rich traditions. No stone is left unturned. The work is slow and systematic. Data banks and the Internet are sieved, all the information is recorded, studied and synthesized using charts and tables. The reference point are the three previous projects that all focus in different ways on interstitial space. They are analyzed, looking for answers that are present but not yet obvious, or not fully comprehensive. An attempt is made to translate the analysis into the input for a design.

1. *Montreal Canadian Exhibition*: inside the wall, through the wall, space is not defined in traditional terms, but instead identifies an interstitial condition.

2. *Milano Triennale*: man's occupation of space and the solid occupation of space. The body resumes an active function.

3. *Church in Rome*: interstitiality and reinforced concrete structures, the formal image as a suggestion.

The latter is the project that appears to have the greatest influence on the new building; in this case, a four-bar struc-

LIQUID CRYSTALS

Liquid crystals are a material phase halfway between liquid and crystal. The molecules are rod-shaped and their order is a function of temperature. The "nematic" phase is characterized by an organization of the constituent molecules that can be controlled through the use of electric fields.

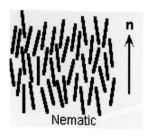

Nematic

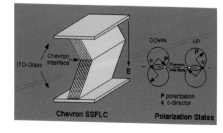

Chevron SSFLC   Polarization States

The more precise condition of being between *in nature is the condition of the liquid crystal, which is a state of suspension between the static crystal and the liquid state. It is the gradual distortion of an originally crystalline phase into a kinematic state, which represents an intermediate phase of the molecular order, before reaching the isotropic or liquid state. The diagrams also show another aspect of liquid crystals, namely that of the layers and multiple superimpositions, seen as deformations of many different layers.*

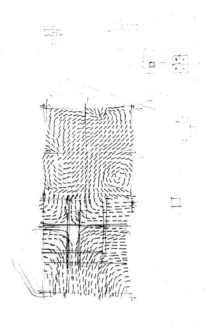

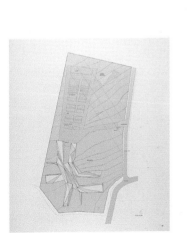

ture replaces the two-bar type used for the church. The bars are not parallel, but are superimposed orthogonally and the deformation system used is again that of liquid crystals.

The study of traditions guides the process of superimposing diagrams; it is true that Eisenman does not always need to link himself directly to the place, but his insertion is always guided by the memory present in every site. Architecture suggests a new way of interpreting, an act capable of describing the nature of the place, revealing its hidden relations through "a sort of atemporal archaeology of urban forms and meanings" (Ciorra 89). Eisenman's interventions try to identify a new approach after using the superimposition of urban traces. This attempt tries to release energy and implications in the layout of the site. The work is therefore based on a number of key points.

Context: The first stage consists of a detailed study of the city of Bangalore in geographical, physical and cultural terms. All the necessary drawings of the project area are collected. Lastly, all the requisites for the program are defined.

Values: Material is collected on the subject of Indian architecture and its traditions; Vastu Shastra is the ancient book of traditional Hindu architecture, *mandala* is a grid used to organise the building, the autochthonous figures range from wells to wide steps. By understanding the implications of Indian architecture, it is possible during the design phase to comply with and use these data as useful values.

Substance of the program: the relations between all the components of the program form the last step that contributes to the success of the building. Its spatial complexity relies on the strength of the program; the building stems from the fusion between the client's requirements and assiduous and constant collaboration with the architect.

The meetings

1. All the material is ready. Every diagram and image are pinned to the wall using a criterion of parallel development between *place* and *project*. For the moment, the latter is only represented by functional schemes derived from the pro-

gram. We start by using the *mandala* to find the right position of the building on the chosen site, before transferring it to the program. Eisenman carefully listens to his collaborators' views and discusses possible changes.

> Something is not quite right yet and we must concentrate on the physical relationships between the spaces. We must act without forgetting the rules of Indian architecture, which create relationships between man, the site and the architectural object. The grid generated by mandala will then be deformed using the conceptual liquid crystal diagrams. The project has still not undergone any deformation because the action is focused exclusively on functional diagrams (Eis 13).

2. All the diagrams continue to be analysed separately, without losing sight of the overall process, as if the superimposition took place outside the image of the architectural object. Each diagram is continuously updated and corrected, there is only a hint of the contact between the various systems of tracings (tradition-program-concept). We proceed in parts, but there is a clear understanding that everything will soon be superimposed, that the project will emerge from the comparison and integration of data. The program has now achieved excellent definition. The main relationships are ready, the functional areas have been identified, the abacus of functional systems and subsystems is now taking shape.

The final solution
The value of the architectural object is not the end result, but the itinerary by which it has been generated. It is important that it emerge as an indissoluble element.
As was said earlier, the project must represent the dualism of Indian society; the strong contrast between the public and private spheres is an example of the interstitial space between two different states. Tradition is represented by a reading of *Vastu-Shastra* and technology is present through the molecular condition of liquid crystals. *Vastu-Shastra* generates a grid on which to position the building, whilst the liquid crystals deform the grid and also shape the building. In conceptual terms, the

BFL Software Limited Bangalore, India

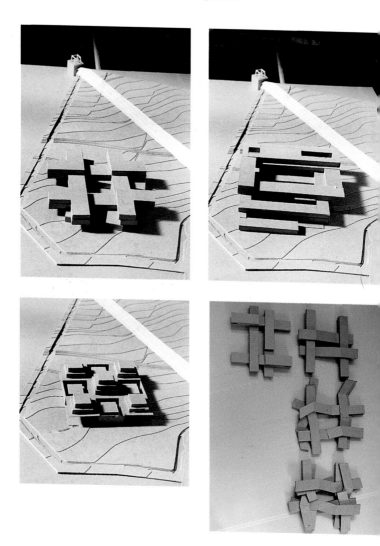

Study models using different distributive schemes.

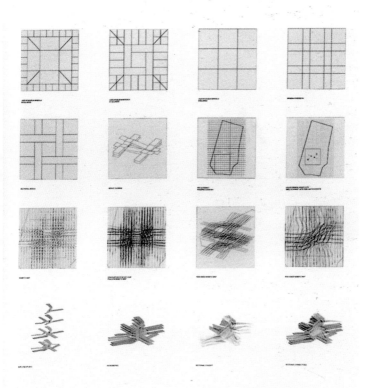

Diagrams.

project tries to represent a software company at the end of the 20th century, but without breaking with tradition. A nine-square *mandala* is laid on the site and is transformed to guide the irregular orientation of the area. The lines of the *mandala* are conceptualized as virtual forces or lines altering the layout of the crystal molecules. This enables the organization of a three-dimensional system of lines used as a dynamic guide for the spaces that were sized and organized earlier using a layout based on parallel bars. The diagrams show a moving geometric form whose characteristics do not comply with Cartesian co-ordinates, but rather a vectorial system, thus creating a kinematic effect. Therefore, in order to understand the movement and to come into contact with space, the onlooker need not move; all that is required is his physical presence.

The building has an area of approximately twelve thousand square meters and returns to the scheme invented for the church in Rome, but with a few essential alterations. Above all, functional changes. The design used for the church was a public space capable of making the visitor feel involved through its movement. The two bars ran parallel and identified the central space as the heart of the building. Here the bars increase and, by crossing over each other, they create a central void that ceases to be a public open void after deformation, turning into an empty space stratified on several layers. This space contains a system of superimposed atriums, linked together by double height ceilings facing the office areas and laboratories housed in the arms of the building.

The four main nodes of the starting scheme, positioned on the different intersections of the bars, are compressed into a single node, from which all the principal spatial units diverge and depart. The research focuses on a space whose function is more clearly defined, overcoming the ambiguities of the voids in the church overlooking the city. The attempt to explore the between is legitimized anew and, through compression, the space is dilated, clarifying the functional distribution; as in the Aronoff Center, the central void sorts the overlapping passages, but here it is not lost in horizontal expansion, but instead clarified through vertical dilatation.

The surfaces do not dematerialize like in the church, but remain compact diaphragms closed to the outside. The outline project is at long last ready.

Instruments for electronic architecture

I use the term "electronic architecture" to define a trend or, more precisely, the experiments achieved in architecture using electronic techniques. This architecture overcomes the barriers of representation and tries to define new parameters of development capable of helping the evolution of representational techniques through a parallel evolution of thought. One of the goals of contemporary architecture is to resolve the conflict between rigid structural organization and freedom, between rules and fluctuating creativity. It is precisely these rules that are used as an instrument to determine the extremely rapid evolution of language. The rules provided by the computer are both free and rigid, depending on how they are used. Understanding a new digital language is only the first step in starting to use it; everyone will find the next one. At present, the users belong to a generation that can only guess at the possible developments without managing them like a real system of thought; but the race has begun. At the dawn of the Computer-Aided Design era, from the Sixties to the early Eighties, CAD systems were expensive, hierarchical and centralized; they played a support role for routine operations performed by large design companies. The software generated construction drawings and standard engineering analyses by rapidly and automatically executing a series of standardized tasks. The arrival of personal computers in the Eighties marked the start of a revolution, and consequently a new use of these machines. Designed as cheap, compact machines, capable of working independently and controlled by a single operator, it was obvious that PCs would become widespread, resulting in the situation as we know it today. There are now a few virtual firms of architects where groups of designers, living in different corners of the globe, share information and exper-

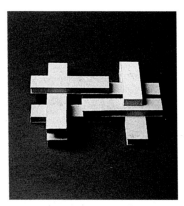
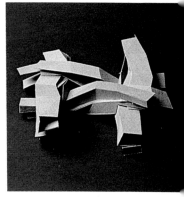

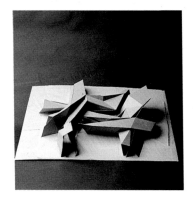

BFL Software Limited, Bangalore, India: model of the 4-bar system, development.

Image showing interior, computer study models.

tise, collaborating on a variety of projects. The computer's influence not only includes the creative process; electronics also affects the future of buildings and the city, it molds their shape and manages space in a functional manner. Buildings are transformed into a system of fields of influence that are flexibly integrated with each other and the cities in their physical and virtual containers.

Faced with this still unexplored creative potential, we must be able to identify the boundary that focuses on the concrete and conceptual limits. The computer allows us to create-manage-think-design; the main question concerns the way of working. Until now, these procedures have been managed separately and creative development has been limited to interacting with the phases of the human mind. Eisenman sees the computer as one of the *starting points* from which to generate constantly changing forms, a partial use that allows creative development to be activated solely with regard to a given portion of the intervention. Instead, it is to be hoped that there will be integration and interference between these "electronic actions" so that architecture can proceed towards its real evolution, released from the stylistic elements of the past.

The notions of space and time have also been influenced by the arrival of electronics and are now drastically compressed: far from being a merely technological question, digitalization has implied a difference in cultural reorganisation. For this to happen, it is vital to adopt the attitude of being willing to undertake this reorganization. You must be willing to support the creation of a different world, one that never existed before. In all this, architecture plays a key role: its status as an art form that follows the laws of the market and can undertake both creativity and innovation enables it not to exclude anything or anyone in advance.

In the electronic era, not only has the character of architecture been altered by the computer, but also its mandate. The traditional view, whereby a set of lines was tidily composed on a sheet has been superseded, and we must now find a new way of thinking, a way to develop and extend an idea by interacting with the digital system. You design space and move around inside it before it is even built: this is the hallmark of a new way of participating. The flexibility and lack of scale typical of "computer architecture" are not the familiar slogans of a rational architectural procedure, but the basic condition for a reinvention of space. The stage is set for unintentional architecture: instead of shapes, architects can abandon terms and rules and turn to data handling, the rebirth of topological architecture. No more boxes and columns: enough of this static nature, here is a space agitated by nothing static, where space must be reappropriated through a purely dynamic vision.

At a conference held in Rome a few years ago, Peter Eisenman shocked the audience by affirming that during the design process he froze the image that seemed to be least comprehensible, and his project then developed from it. This provocative statement, which is not so difficult to interpret if you look beyond the material relations with the machine, served as the legitimization of a starting point, the identification of a line to follow.

The machine alters objective reality by changing it into new realities that are completely different from the known world. You enter a completely new field where new horizons open up to the imagination. The computer is a machine that makes transformations, and the idea of architectural space or, more generically, architecture, lends itself to this. Of the four basic transformations performed by the computer – translation, atomisation, logicisation and metaphorisation – the latter represents the creative act. While the computer works through a constant effort of formal transposition, we can correlate different data and contents. The architecture of diagrams is correlated with the vectorial system, and their materialized forms provide the basis for rewriting the architecture of the third millennium.

Diagrams

Derrick de Kerckhove affirms that it is important to understand the absolute value of the computer which, like the

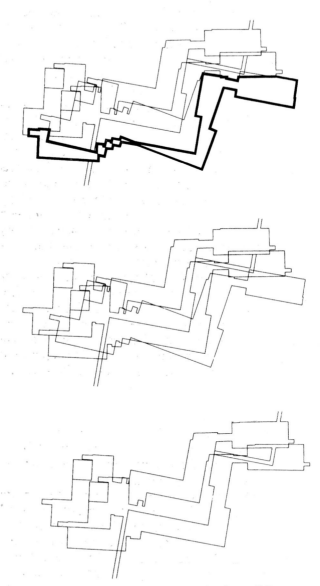

Aronoff Center, diagrams.

alphabet, is completely independent of meaning and must be used to transform a sequence of data. These transformations parallel the development of the project. The translation of information occurs through a metaphorization process that helps us to express the primary concepts adequately.

Eisenman uses this system at the start of the process, while the diagrams prompt reflection and provide a series of possible paths to follow. The computer elaborates and accelerates thought, ideas lose their materiality and become information. Marshall McLuhan guessed the final goal of the mutation produced by the computer. Eisenman records it, activates its infinite series of mutations and integrates them with information from other disciplines. Mathematics, physics and biology become information and concepts that can be transformed into architectural concepts. The American architect's theory becomes a window on contemporary society, and the computer structures and manages this information system.

In the project for the Aronoff Center, for the first time ever Eisenman started the electronic management from an idea: the diagrams that had always accompanied his projects become more complex. The techniques used in this construction achieve a dual effect: they combine specific architectural parameters in unconventional ways and reject the application of external models that imply negotiations with the spatiality of the spirit of the time.

Breakdown of the building

The first part of the process is guided by geometric and functional diagrams, divided into: 1. Functional diagrams; 2. Interpretation of the context; 3. Superimpositions; 4. Deformation; 5. Twisting; 6. Superimposition; 7. Shift.

The second part of the diagrams is expanded from the system of pre-existing forms. The dimensions of the profiles are used to extend a similar figure through the site, thus connecting the buildings together and with the new building. This profile is rotated and translated so that it is perpendicular to first one and then the other existing building. The space is identified between the profiles and the figures.

The plan of the building starts to look like a bicycle gear that does not move mechanically, but whose movement results from the articulation of the entire figure. It would be impossible to follow the construction and measure these spaces using traditional methods. Any section of the building is perpendicular to only one of the overlapping boxes. This means that the use of the computer is indispensable.

The system of co-ordinates is used to measure spaces. Every point can be identified using spatial co-ordinates on Cartesian axes x-y-z from a source point identified in the old building. The contractor has used a laser system to make the measurements. The building is positioned and constructed using the same system.

The interesting thing about this method is that it manages to release the project from the idea of architecture as a traditional system and to suggest the possibility of architectural knowledge through pure production. Only by doing so is it possible to overcome what, following a superficial analysis, might appear to be a pointless formality. This is an extremely fascinating and innovative method for a building that, lacking any set fixed points, can alter at will, following its own outlines.

Axes and vectors

> We can draw an axis from the mind to the hand, thanks to our knowledge of the human body, but we can use the computer to represent a vector, which has nothing to do with an axis. A vector has density, a direction, a force that we cannot draw. We cannot conceptualise a vector, but computers can [...]. We open up a completely new world of possible architectural expressions and experiments (Eis 13).

In today's accelerated technological context, we must force ourselves to make a real effort to integrate or, better still, accept all the devices that we have built, immediately realizing that every instrument will undermine our psyche and become our new extension, what De Kerckhove describes as a phantom limb, one which we can feel but cannot move.

Having reached the end of this initial phase of consternation, technological art, the ability to manage our extensions will reach a new frontier. Our modern technologies are so versatile that they give us the power to redraw reality, therefore making it necessary to abandon our one-dimensional points of view and recognize that they are rapidly overcome by the new perception of our "status point," the only physical reference point in the total winding of our electronic projections.

We must overcome our typical visual experience, replacing it by a tactile, emotional experience, which contains a strong realistic connotation based on body sensations, what Eisenman calls "destabilization" or, better still, "body involvement" in his architecture. To return to De Kerckhove, the "status point" replaces the point of view understood as distance and becomes the point of entry to a new reality.

Only our old visual prejudices can still prevent us from seeking what is obvious: interactivity is contact, all the meanings of technological extensions must be rethought, but must not be regarded as simple aids to our design, but rather forms and systems of thought capable of generating space through simulation.

But what can the computer do? How can it guide the project? First of all, the computer can give us an accurate picture of the complexity of moving forms.

Plans and elevations are replaced by an endless sequence of views and solutions that materialize ideas and abstract concepts, making them take part in the complexity of a space. The essential process of this era consists in the dissolution of objects: they lose their materiality and are transformed into information that, having been processed, is translated into a new form of materiality, new objects. The computer can change the way we think, it will soon think with us. This change will also influence the space in which we live; this continuous re-elaboration will make us lose sight of the original information, indeed we will doubt that it ever existed. In the passage from the mind to the hand, we are in a sense obliged to draw axes; using the computer, it is possible to draw a vector, which has nothing to do with an axis. Vectors have density and precise properties, they can express concepts and

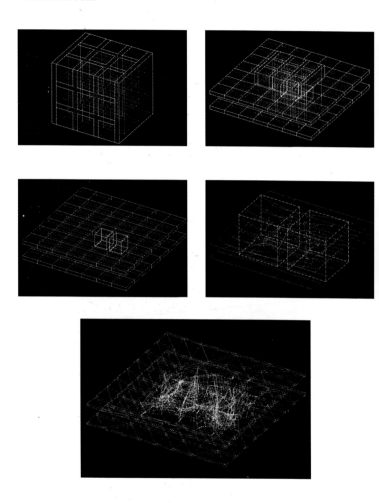

Genesis of the project for a virtual house. The house originates from the interaction of nine cubes which constitute a potential field of vectorial relationships. Each vector deforms a grid line, the lines themselves become forces in movement which interact with one another.

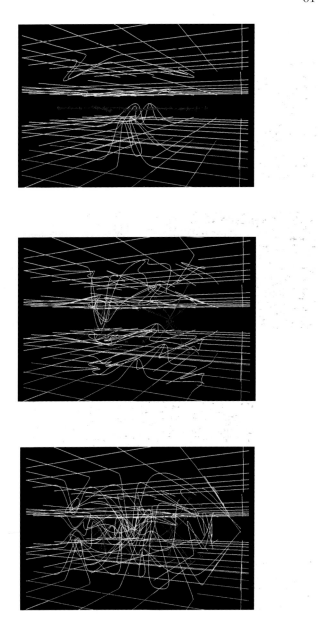

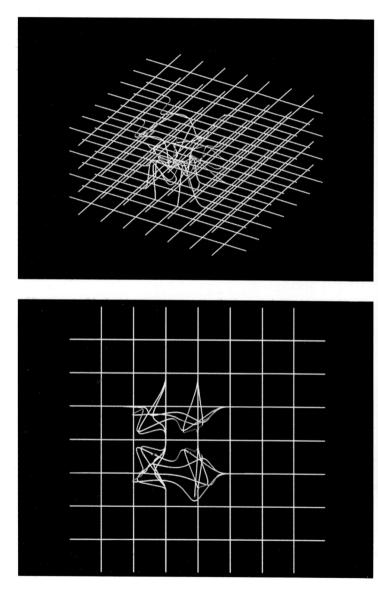

Virtual House, *diagram of genesis. From the movement and the relation-ship between the vectors, a completely new spatial condition is created,*

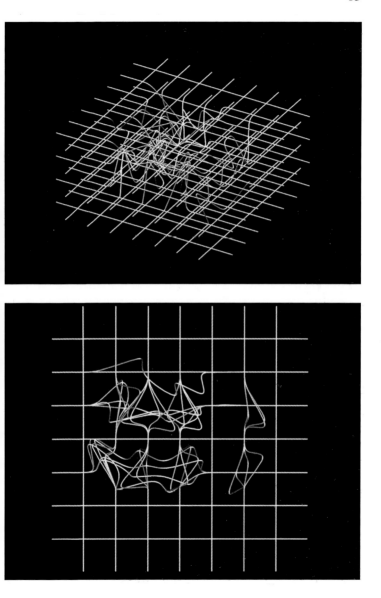

with infinite variations. The layout of the lines in movement becomes the physical limit of the space in movement.

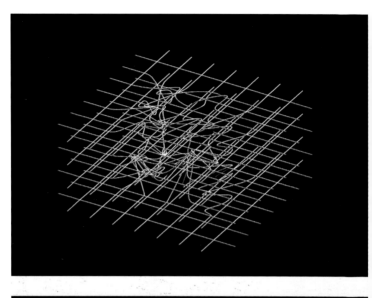

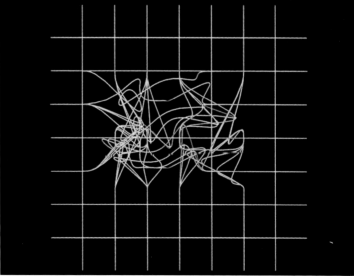

Virtual House, *diagram of genesis. Page 65: diagrammatic model. The process could infinitely repeat itself, but the number of repetitions has been assigned "a priori." The machine created this way slows down when the*

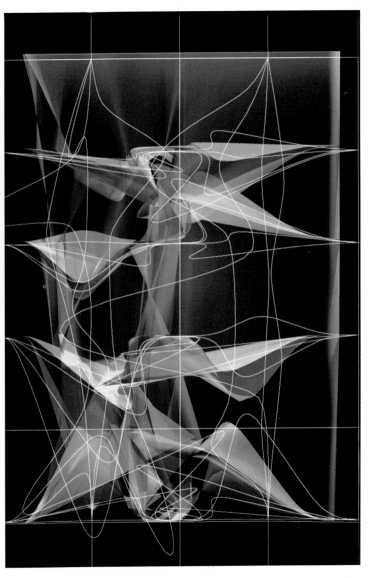

influence between the outlines is reduced to a minimum. At this point, the surfaces are outlined between the lines. The space between these surfaces is an architectural space to be discovered using the operating program.

model space. A vector has a direction, a force that we cannot draw. In a drawing we can only represent something that is known, whereas the formal potential of the vector is infinite and cannot be controlled using a set process.

By working with computers, we have opened up a new chapter in architectural experimentation. The machine transcribes vectors, temporally updating their parameters, binding them and achieving unexpected expressive goals. The design for a virtual house, for example, shows a field of variable intensity as the project site. The condition of the field is an assembly of points and lines, whose edges are the result of an action-reaction of internal constraints produced by the machine. These conditions produce the figure in relation to vectors and time.

> But it is still necessary to think and see in three dimensions, given that architecture in the era of the media and images must give responses of effective spatiality and corporeality in terms of space (Eis 14).

At an everyday level of design, a dialogue is established between two different models of project development; 3D models are constantly produced, but always after conceptualization, which takes place at the computer in a process of continuous refinement. The computer becomes the materialization of thought, which is still far from any architectural realization, but needs to be visualized; at this point, the model is created through suggestion. The design is always completed after having seen 3D models. As was mentioned earlier, vectors appear for the first time in the Virtual House.

Virtual House

In this way form, or rather "non-form" becomes the natural expression of virtual reality. Work on the *Virtual House* was based on the memory of the spatial concepts for a house for which Eisenman wrote a text in 1987, entitled *Virtual House*. The house initially stemmed from the interaction of nine cubes that constitute a potential field of internal relations and interconnections.

Each connection can be expressed as a vector; a field of influence is attributed to each vector, which updates its virtual movement through time. This updating is visualized by the effect of each individual vector on a line. The lines, together with their geometric properties, become moving forces. These correlated movements produce a mechanical system capable of generating forms influenced by the position and orientation of the vectors. The tracing left by these moving lines becomes the boundaries of space.

Two distinct phases can be identified in the genesis of the project. In the first phase, two of the original nine cubes are isolated with sides next to one another. The computer reads each side from corner to corner and the tracing of this reading is recorded. This process is repeated again, this time with the two cubes constrained by one another; the tracing is recorded again. The process could be repeated several times using a number of cubes. The number of cubes (two) and the number of times (two) was chosen a priori for this project. The machine slows down when the movement reaches a stage of reduced activity in each repetition, namely when the differences are no longer perceptible. Using this system, time absorbs the formal expression of the virtual. We see the outline of the shape appearing, the functional spaces beginning to articulate themselves inside it.

> The virtual constitutes the entity: the virtualities inherent in a being, its problems, the node of tensions, constraints and projects that brings it to life, the questions that give it movement, are an essential part of its determination (Eis 14).

The house materializes, and suddenly the traces used to generate it disappear and several of Eisenman's projects appear in the background; it is like watching a hologram, the church with its two building volumes rising out of the ground, the curves modelling the space of the Aronoff Center passing through the modelling of the site in Rebstock Park. The ground is no longer a slab on which the volume rests, but becomes a compact fabric that forms the house itself; a system in which the functions and various habitable spaces inter-

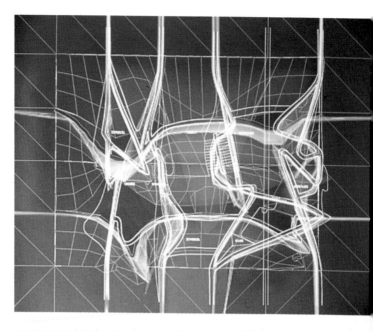

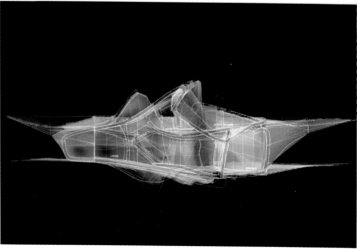

First level floor plan and longitudinal section. Page 69: second level floor plan and longitudinal section. Following the diagrammatical genesis, there

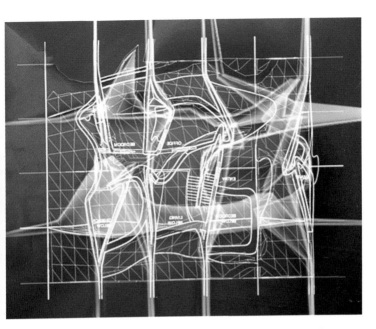

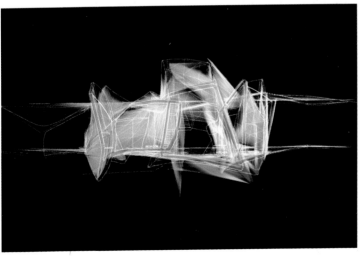

is a functional investigation of the spaces discovered through vectorial deformation.

act "to oppose discontinuity using the continuity between one space and the next" (Saggio 96). New spaces open and close, one inside the other, depending on their use, defining new possibilities for living. The light penetrates through the folds of the skin and at the points where the membrane divides in two. Access is through the upper part, sliding inside, where Eisenman again tries to destabilize the usual spatial references using a gravity-free architecture. The most fascinating element is the possible position of the house, which can be built above or below ground level because it is generated between two parallel lines, without detracting in the slightest from living conditions.

Like Narcissus, Eisenman sees his own reflection within himself and generates a project that he himself did not expect; the forms contract and seek concrete spatiality through movement, where at long last "the virtual acquires full reality through being virtual" (Deleuze).

The new system of vision

During the fifty years since the Second World War, a paradigm shift has taken place that should have profoundly affected architecture: this was the shift from the mechanical paradigm to the electronic one. This change can be simply understood by comparing the impact of the role of the human subject on such primary modes of reproduction as the photograph and the fax: the photograph within the mechanical paradigm, the fax within the electronic one (Eis 14).

Vision is the system of relations used to define the typological conditions of architecture. It is the way of defining a relationship between a subject and an object, a way of organizing space and spatial elements. Vision today is still the focal point of our attention, but the subject is no longer man, his interaction with the object is devalued, the possibility of modifying and participating has ended because our vision is no longer interpretative. The electronic paradigm devalues the original, it produces a copy, creating new parameters, a new way of reading and interpreting reality. However, contrary to expectations, this has not yet led to major changes in architecture. Architecture has always been associated with reality, with what exists and is present in a totally physical and concrete objectiveness; the electronic paradigm represents a powerful challenge to architecture.

> Because it defines reality in terms of media and simulation, it values appearance over existence, what can be seen over what is. The media introduce fundamental ambiguities into how and what we see (Eis 14).

The mechanism of vision shapes architecture and its representation, but is still linked to the perspective which was elaborated in the fifteenth century. This extremely antiquated theory of space allows for all projections of space to be resolved on a single planimetric surface. It is this possibility of representing architectural space on a two-dimensional surface that has had such a profound influence on architectural practice over the years, imposing a spatial organization based on axes, points and symmetries, in order to orient the body through the eye and reinforce the centrality of the observer. Contrary to what one might believe, there is nothing natural about perspective: it is an artificial way of drawing. Children do not worry about drawing objects in perspective and many civilizations, for example the Egyptians, Chinese or African cultures, have been broadly speaking indifferent to the use of perspective.

Computers are accelerating our psychological response and our reaction times. It is now possible to observe 3D models that were previously impossible to check so rapidly. The design response, or rather the initial spatial verification, almost occurs before the real model studied on different scales. The sketch has been replaced by a precise image which suggests the development of the entire project thanks to the reading of these electronically generated emotional spaces. The order has been inverted, the physical model is anticipated by the virtual which possesses intuitive value.

The new architecture is both inside and outside the computer, it is made up of communication networks that use cables, optic fibers, tributaries of the same technology: electricity is the only new, common language. The architect's work based on the new media and new machines does not consist in praising or condemning technology, but overcoming the hiatus between technology and psychology. There is an extraordinary possibility of seeing more beyond the existing boundaries; as De Kerckhove writes, the secret is to see behind our back, with total involvement.

It is necessary to rethink and work along the lines of Piranesi, the only modern architect capable of shifting vision through a new use of perspective; equally, Cubism attempted to alter the relationship between subject and the object, in a hypnotic overturning capable of displacing our vision. But architecture has not succeeded in doing the same, the subject is still totally immersed in an anthropocentric vision. The object has not succeeded in displacing the subject-onlooker, and architecture has not overturned the concept of vision because, as Eisenman affirms, it has embodied the vision and at the same time because it is the only discipline that permits man to be effectively inside or outside, to move around and control his relation with the object. What is now needed is an upheaval that will change the subject's role, separating it from the rationalization of space. To do this, the electronic paradigm will help us to overthrow normal construction by separating the eye from the mind, thus acting directly on the feelings. Our eyes are now incapable of controlling reality.

> The computer gives you the possibility of constructing objects that you would never do directly from the mind to the hand [...] It is still necessary to think, to see in three dimensions because architecture in the era of the media and images must respond with effective spatiality and corporeality in terms of space. We constantly produce models after having conceptualized them using the computer. It is a process of constant refinement (Eis 14).

Eisenman suggests looking for the other space by detaching what one sees from what one knows. He suggests re-writing space, even if we are used to seeing it written using traditional precepts and elements: windows, columns and decorations. Re-elaborating the inscription of space means overturning it – not mechanically, but conceptually, using new instruments – without that design intention that does nothing but reconstruct the vision in the terms we wish to overcome. Eisenman looks for an instrument that will allow him to break the established patterns, seeking an unbroken continuity between inside and outside, namely overcoming the pre-established system of registration: he finds it in Deleuze's folds.

Folded space articulates a new relationship between vertical and horizontal, figure and ground, inside and out. The idea of folded space denies framing in favour of a temporal modulation. The fold no longer privileges planimetric projection; instead there is a variable curvature (Eis 13).

How can we define this variable curve except using a method of representation or, again, a new and dynamic experimental method? In the digital era, more than any other, the computer represents the most suitable instrument for displacing traditional perspective vision, thus changing the machine from a technical aid into a subject that generates a destabilizing action. The digital instrument can be used to show reality as never before.

Eisenman understands an essential element and tries to exploit its abstract possibilities. It is a fluid architecture, where shapes become data fields, an architecture that resembles the urge to rewrite our reality, used by Eisenman to focus on the point of a new avantgarde performance. This experimentation stems from the desire to create a space that breaks through the boundaries between real and virtual, in a discourse whose words are still partly inside the computer.

Eisenman's process continues to change and evolve. The traces of the past are increasingly evident in the most recent projects, what changes is the use of technology. Eisenman looks for an answer to his endless questions using digital systems, he does not overturn his architecture, but lets himself be overturned in order to accelerate his way of creating space.

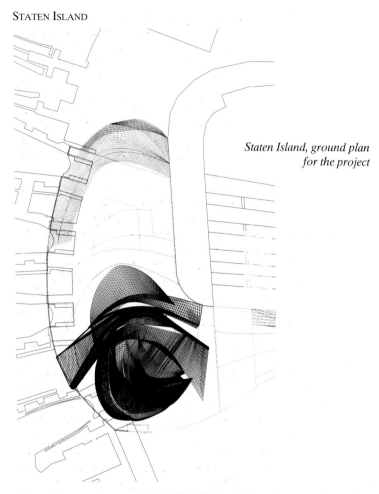

*Staten Island, ground plan
for the project*

The software that could fulfill all his wishes does not yet exist, but his research goes on. His diagrams are corrected by hand and even the projects change after they are modelled on paper, thanks to the computer's suggestions. This means that Eisenman does not appear to be consistent; all too often his discourses or, more still his texts, do not coincide with his architecture, but then the diagrams revise everything, it all coincides, the process is concluded and no one stage prevails over another. The capacity to be inconsistent during the process leads him in the end to an extremely consistent project, one that sets the new rules of the game once and for all.

Taking the computer as their starting point, too many architects today abandon orthogonal lines and seek refuge in other geometric forms, which are always present in the spaces in which we live. Their revelation is almost a requisite; dictated by the new digital paradigms. The complexity lies in the main reference systems with which it is possible to distinguish between the two types of spatial layout, *effective* and *affective*. The first includes movements, figures and activities in a broader organization that anticipates and survives them; in the second an attempt is made to create figures or movements from the various spatial organizations following unexpected itineraries that are loosely related. Moreover, constructions and intuitions acquire a different meaning in the two cases: in the first an attempt is made to draw all the lines of our various geometric figures starting from the individual fixed points of an a priori system, whilst the second works on a more informal and arbitrary theoretical diagram.

We can therefore talk about two types of geometry: one that is prearranged and another that is intuitive and topological. These reveal two different tendencies. On the one hand, an abstract tendency that tries to build logical relations based on the varied materials available, therefore ordering and linking them in a systematic fashion; on the other, an intuitive tendency that searches for a clear perception of the objects examined and a concrete representation of their reciprocal relations. Eisenman throws himself into this process with unexpected energy. His spaces emotionally involve man, who is obliged to reinvent his own references, and this invention is prearranged at every stage of the process and modified by the computer that amplifies the attempt at destabilization.

In the computer I look for the operating conditions, I look for what I cannot understand, or rather something that fascinates me because it is unfamiliar (Eis 5).

Eisenman is not content to amaze and fascinate, he wants to convince us about a method, he wants to assure us that behind every project there is a precise construction without which architecture would be impossible.

The stream of data and digital models is not alone sufficient to justify or trigger off an architectural process capable of shaping everyday space. Eisenman seeks to organize the electronic input systematically through other disciplines that use the electronic era and its media as thinking tools. Computer, philosophy, science and biology interact to reinvent space. In the study of architecture, the computer becomes the ideal instrument with which to change the relationship between the drawing of the project and real space.

Digital thought can therefore trigger off an evolutionary process of the project without any prearranged meaning. Folding, for example, is the start of a journey, an intuition that others have had but have not managed to explore using the appropriate means, always staying inside Cartesian space. Eisenman overcomes this by enabling the subject to take part in a different way, returning to a primitive logic dictated by emotion.

We should not stop at traditional interactive criteria, but must look beyond: information technology helps us to enter into and shape the new space. Dislocating vision is both possible and necessary, and folding is only one of the possible ways of responding to the Cartesian order. Drawing no longer has any scale value relationship in the three dimensional environment, whereas digital modelling interprets space by suggesting ideas instead of certainties, logic instead of vision; it determines a dislocation of the dialectic distinction between figure and plane. This leads to the creation of affective space; architectural projects will always have four walls, but they will no longer symbolize the mechanical paradigm, but rather the electronic one.

The dislocation of the observer is not achieved through the traditional system, working on the translation of the same axis, but rather through a stratified system that constantly shatters before being recomposed on the diagram of flows and vectors. To use Eisenman's own words: spatial experience is that offered by an unfinished tale, an endless journey.

In the Staten Island Institute of Arts and Sciences, the central 74-83 reference axis is transformed by twisting a dense mass. At first sight, the spatial invention used by Eisenman in this building recalls the deformations of the better known Museum in Bilbao designed by his friend, Gehry. But there are significant differences. Gehry does not stray from tradition and organizes his museum around vertical and horizontal axes; the final composition is a static one, even if apparently new, an architecture that still requires an axis.

In Bilbao, movement is frozen through the superimposition of the individual objects, whereas in Staten Island, Eisenman creates movement through a stratification that is unrelated to any continuous axis. The axis is broken by a vectorial system that organizes the layers by superimposing them. Looking at the diagrams, we can identify the deformation of the grids used by the individual elements correlated by the dynamic nature of flows. The computer accelerates thought and guides it along a path whose end is out of sight. It is not a question of creating architecture, but merely guiding it towards intuitions that would otherwise be uncontrollable.

The non-intentional nature of Eisenman's actions is only a necessary condition, turning the chain of events that determine its development into a "machinic process," namely a set that operates independently of the meaning of its elements.

Each project is a different process and the next is realized on the basis of the previous one; the use of the computer completes and enhances each of these processes and lays the foundations for future development. In fact, only through experience will it be possible to develop the capacity of the instruments, leading to constant collaboration and improvisation with other players.

To use Zaera Polo's description, Eisenman is the first real "machinic architect" capable of destabilizing or deviating from many of the conditions of meaning immanent since the beginning, transforming them into completely new ideas.

Notes on Staten Island

by Marco Galofaro

It is never easy to write about an architect's work. Doing so in relation to a project on which you have yourself worked should be easier, but not when the project is by Peter Eisenman. It is important to separate one's own judgement from the experience as a whole, to remove oneself from the center of action in order to be objective, to work critically on something of which a small part belongs to the sphere of one's own emotions. Only a few months ago, I worked in Eisenman's office on the Staten Island Institute of Art and Sciences: the object in question is in my mind, in reality, in the same way as when you draw a sketch (of a landscape or a face), and it remains in my memory when it becomes the outline of a given project-object-image. It takes time to organize one's own ideas, one's own evaluations, time is essential for the purity of the argument, so that nothing appears obscure when it is developed over time and according to its own laws (Shakespeare wrote of the dramatic need to talk of the present in unchanging terms or to be silent for ever). I now find myself writing, encouraged by my brother, far away from the office in 25th Street, but with a more vivid and perhaps more conscious memory. An encounter with Eisenman's architecture is something that inevitably prompts reflection: his architecture represents breakdown, a state of crisis, change. Staten Island Institute is no different, it inevitably prompts reflection, it is an architecture of pure form, without wishing to understand the expression in the restricted sense in which it is often used. Pure form that goes beyond representation, continuing a research that has always tried to extract architecture from its own matrix of meaning.

The project for Staten Island is a ferry terminal, a nodal point for bus, car and ferry transport, used by over eighteen million persons every year. Its movement is its dominant trait. Eisenman transforms this movement into a fluid object, the flow of itineraries molds the space, the form envelops and twists on a central void, a theme that is common to many projects of this kind, but the anti-classic fury of this architecture (even though based on a conventional scheme) leads it to deny the large central axis because no continuous itinerary links the parts, structured in alternating layers of solid and void. The building takes shape within this dilated and compressed space, offering the users of the terminal a series of services, restaurants, an Imax theatre, shops, teaching facilities and exhibition areas. The waiting room for the terminal becomes the museum hall. The itineraries blend between the flows of people moving through them. The architectural object is not recognizable per se; its dissolution into bands intersected by vibrant slits and deep V-shaped grooves indicate the presence of an active, magmatic underground energy. The earth shakes, undermining the ground's solidity.

His theoretical obsession, focused as always on the breakdown of every absolute value of rationalist origin, clashes in this project with what has always been the key support of architecture: the ground. No longer a static ground, a simple presence, but a telluric stratification that interacts with the whirling spatial process characteristic of the project. No part is formally connotated, to the point that it is impossible to identify the entrance itself. The sense of the building is totally denied. All this inevitably leads me to a reflection on the dual nature of Eisenman's work. His work violates and betrays tradition; perhaps because this is the only way of preserving it? This means that there is a revolutionary Eisenman and a conservative Eisenman. But this dual tension always leads us to look for a new interpretation, in an attempt to discover, to penetrate new areas of meaning.

The forms of Eisenman's architecture can only exist if immersed in their evolution. The nature of space can only be dynamic, fluid, telluric because its fundamental characteristic is mobility.

Staten Island, project.

Staten Island, project.

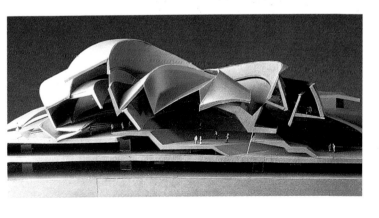

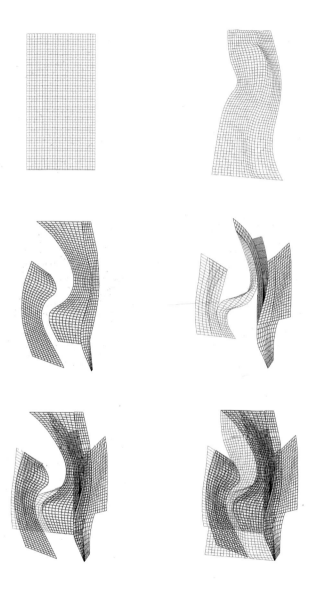

Staten Island, diagrams.

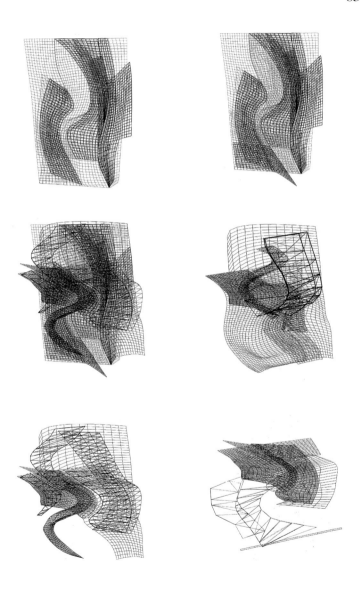

Staten Island, diagrams.

Visions unfolding: architecture in the age of electronic media

by Peter Eisenman

During the fifty years since the Second World War, a paradigm shift has taken place that should have profoundly affected architecture: this was the shift from the mechanical paradigm to the electronic one. This change can be simply understood by comparing the impact of the role of the human subject on such primary modes of reproduction as the photograph and the fax; the photograph within the mechanical paradigm, the fax within the electronic one.

In photographic reproduction, the subject still maintains a controlled interaction with the object. A photograph can be developed with more or less contrast, texture or clarity. The photograph can be said to remain in the control of human vision. The human subject thus retains its function as interpreter, as discursive function. With the fax, the subject is no longer called upon to interpret, for reproduction takes place without any control or adjustment. The fax also challenges the concept of originality. While in a photograph the original reproduction still retains a privileged value, in facsimile transmission the original remains intact but with no differentiating value since it is no longer sent. The mutual devaluation of both original and copy is not the only transformation affected by the electronic paradigm. The entire nature of what we have come to know as the reality of our world has been called into question by the invasion of media into everyday life. For reality always demanded that our vision be interpretive.

How have these developments affected architecture? Since architecture has traditionally housed value as well as fact, one would imagine that architecture would have been greatly transformed. But this is not the case, for architecture seems little changed at all. This in itself ought to warrant investigation, since architecture has traditionally been a bastion of what is considered to be the real. Metaphors such as house and home; bricks and mortar; foundations and shelter, attest to architecture's role in defining what we consider to be real. Clearly, a change in the everyday concepts of reality should have had some affect on architecture. It did not because the mechanical paradigm was the sine qua non of architecture; architecture was the visible manifestation of the overcoming of natural forces such as gravity and weather by mechanical means. Architecture not only overcame gravity, it was also the monument to that overcoming; it interpreted the value society placed on its vision.

The electronic paradigm directs a powerful challenge to architecture because it defines reality in terms of media and simulation, it values appearance over existence, what can be seen over what is. Not the seen as we formerly knew it, but rather a seeing that can no longer interpret. Media introduce fundamental ambiguities into how and what we see. Architecture has resisted this question because, since the importation and absorption of perspective by architectural space in the

15th century, architecture has been dominated by the mechanics of vision. Thus architecture assumes sight to be preeminent and also in some way natural to its own processes, not a thing to be questioned. It is precisely this traditional concept of sight that the electronic paradigm questions.

Sight is traditionally understood in terms of vision. When I use the term *ision*, I mean that particular characteristic of sight which attaches seeing to thinking, the eye to the mind. In architecture, vision refers to a particular category of perception linked to monocular perspectival vision. The monocular vision of the subject in architecture allows for all projections of space to be resolved on a single planimetric surface. It is therefore not surprising that perspective, with its ability to define and reproduce the perception of depth on a two-dimensional surface, should find architecture a waiting and wanting vehicle. Nor is it surprising that architecture soon began to conform itself to this monocular, rationalizing vision – in its own body. Whatever the style, space was constituted as an understandable construct, organized around spatial elements such as axes, places, symmetries, etc. Perspective is even more virulent in architecture than in painting because of the imperious demands of the eye *and* the body to orient itself in architectural space through processes of rational perspectival ordering. It was thus not without cause that Brunelleschi's invention of *one-point* perspective should correspond to a time when there was a paradigm shift from the theological and theocentric to the anthropomorphic and anthropocentric views of the world. Perspective became the vehicle by which anthropocentric vision crystallized itself in the architecture that followed this shift.

Brunelleschi's projection system, however, was deeper in its effect than all subsequent stylistic change because it confirmed vision as the dominant discourse in architecture from the 16th century to the present. Thus, despite repeated changes in style from the Renaissance through Post Modernism and despite many attempts to the contrary, the seeing human subject – monocular and anthropocentric – remains the primary discursive term of architecture.

The tradition of planimetric projection in architecture persisted unchallenged because it allowed the projection and hence, the understanding of a three-dimensional space in two dimensions. In other disciplines – perhaps since Leibniz and certainly since Sartre there has been a consistent attempt to demonstrate the problematic qualities inherent in vision, but in architecture the sight/mind construct has persisted as the dominant discourse.

In an essay entitled *Scopic Regimes of Modernity*, Martin Jay notes that, "baroque visual experience has a strongly tactile or haptic quality, which prevents it from turning into the absolute ocular centrism of its Cartesian perspectivalist rival." Norman Bryson in his article, *The gaze in the expanded field*, introduces the idea of the gaze (*le regard*) as the looking back of the other. He discusses the gaze in terms of Sartre's intruder in *Being and Nothingness* or in terms of Lacan's concept of a darkness that cuts across the space of sight. Lacan also introduces the idea of a space looking back which he likens to a disturbance of the visual field of reason. From time to time architecture has attempted to overcome its rationalizing vision.

If one takes, for example, the church of San Vitale in Ravenna, one can explain the solitary column almost blocking the entry or the incomplete groin vaulting as an attempt to signal a change from a Pagan to a Christian architecture. Piranesi created similar effects with his architectural projections. Piranesi diffracted the monocular subject by creating perspectival visions with multiple vanishing points so that there was no way of correlating what was seen into a unified whole. Equally, Cubism attempted to deflect the relationship between a monocular subject and the object. The subject could no longer put the painting into some meaningful structure through the use of perspective. Cubism used a non-monocular perspectival condition: it flattened objects to the edges, it upturned objects, it undermined the stability of the picture plane. Architecture attempted similar dislocations through Constructivism and its own, albeit normalizing, version of Cubism – the International Style. But this work only *looked* cubistic and modern, the subject remained rooted in a profound anthropocentric stability, comfortably, upright and in place on a flat, tabular ground. There was no shift in the relationship between the subject and the object. While the object looked different, it failed to displace the viewing subject. Though the buildings were sometimes conceptualized, by axonometric or isometric projection rather than by perspective, no consistent deflection of the subject was carried out. Yet modernist sculpture did in many cases effectuate such a displacement of the subject. These dislocations were fundamental to minimalism: the early work of Robert Morris, Michael Heizer and Robert Smithson. This historical project, however, was never taken up in architecture. The question now begs to be asked: Why did architecture resist developments that were taking place in other disciplines? And further, why has the issue of vision never been properly problematized in architecture?

It might be said that architecture never adequately thought the problem of vision because it remained within the concept of the subject and the four walls. Architecture, unlike any other discipline, concretized vision. The hierarchy inherent in all architectural space begins as a structure for the mind's eye. It is perhaps the idea of interiority as a hierarchy between inside and outside that causes architecture to conceptualize itself ever more comfortably and conservatively in vision. The interiority of architecture more than any other discourse defined a hierarchy of vision articulated by inside and outside. The fact that one is actually both inside and outside at architecture, unlike painting or music, required vision to conceptualize itself in this way. As long as architecture refuses to take up the problem of vision, it will remain within a Renaissance or Classical view of its discourse. Now what would it mean for architecture to take up the problem of vision? Vision can be defined as essentially a way of organizing space and elements in space. It is a way of looking *at*, and defines a relationship between a subject and an object. Traditional architecture is structured so that any position occupied by a subject provides the means for understanding that position in relation to a particular spatial typology, such as a rotunda, a transept crossing, an axis, an entry. Any number of these typological conditions deploy architecture as a screen for looking-at.

The idea of a "looking-back" begins to displace the anthropocentric subject.

Looking back does not require the object to become a subject, that is to anthropo-morphosize the object. Looking back concerns the possibility of detaching the subject from the rationalization of space. In other words, to allow the subject to have a vision of space that no longer can be put together in the normalizing, classicizing or traditional construct of vision; an other space, where in fact the space "looks back" at the subject. A possible first step in conceptualizing this "other" space, would be to detach what one sees from what one knows – the eye from the mind. A second step would be to inscribe space in such a way as to endow it with the possibility of looking back at the subject. All architecture can be said to be already inscribed. Windows, doors, beams and columns are a kind of inscription. These make architecture known, they reinforce vision. Since no space is uninscribed, we do not see a window without relating it to an idea of window; this kind of inscription seems not only natural but also necessary to architecture. In order to have a looking back, it is necessary to rethink the idea of inscription. In the Baroque and Rococo such an inscription was in the plaster decoration that began to obscure the traditional form of functional inscription. This kind of "decorative" inscription was thought too excessive when undefined by function. Architecture tends to resist this form of excess in a way which is unique to the other arts, precisely because of the power and pervasive nature of functional inscription. The anomalous column at San Vitale inscribes space in a way that was at the time foreign to the eye. This is also true of the columns in the staircase at the Wexner Center. However, most of such inscriptions are the result of design intention, the will of an authorial subjective expression which then only reconstitutes vision as before. To dislocate vision might require an inscription which is the result of an outside text which is neither overly determined by design expression or function. But how could such an inscription of an outside text translate into space?

Suppose for a moment that architecture could be conceptualized as a Moebius strip, with an unbroken continuity between interior and exterior. What would this mean for vision? Gilles Deleuze has proposed just such a possible continuity with his idea of the fold. For Deleuze, folded space articulates a new relationship between vertical and horizontal, figure and ground, inside and out – all structures articulated by traditional vision. Unlike the space of classical vision, the idea of folded space denies framing in favor of a temporal modulation. The fold no longer privileges planimetric projection; instead there is a variable curvature. Deleuze's idea of folding is more radical than origami, because it contains no narrative, linear sequence; rather, in terms of traditional vision, it contains a quality of the unseen. Folding changes the traditional space of vision. That is, it can be considered to be effective; it functions, it shelters, it is meaningful; it frames, it is aesthetic. Folding also constitutes a move from *e*ffective to *a*ffective space. Folding is not another subject expressionism, a promiscuity, but rather unfolds in space alongside of its functioning and its meaning in space – it has what might be called an excessive condition or *a*ffect. Folding is a type of *a*ffective space which concerns those aspects that are not associated with the *e*ffective, that are more than reason, meaning and function. In order to change the relationship of perspectival projection to three-dimensional space it is necessary to change the relationship between project drawing and real space. This would mean that one would no longer be able to draw with any level of meaningfulness the space that is being projected. Fox example, when it is no longer possible to draw a line that stands for some scale relationship to another line in space, it has nothing to do with reason, of the connection of the mind to the eye. The deflection from that line in space means that there no longer exists a one-to-one scale correspondence.

My folded projects are a primitive beginning. In them, the subject understands that he or she can no longer conceptualize experience in space in the same way that he or she did in the gridded space. They attempt to provide this dislocation of the subject from effective space; an idea of presentness. Once the environment becomes affective, inscribed with another logic or an ur-logic, one which is no longer translatable into the vision of the mind, then reason becomes detached from vision. While we can still understand space in terms of its function, structure and aesthetic – we are still within the "four walls" – somehow reason becomes detached from the affective condition of the environment itself. This begins to produce an environment that "looks back" – that is, the environment seems to have an order that we can perceive, even though it does not seem to mean anything. It does not seek to be understood in the traditional way of architecture, yet it possesses some sense of "aura," an ur-logic which is the sense of something outside of our vision. Yet one that is not another subjective expression. Folding is only one of perhaps many strategies for dislocating vision – dislocating the hierarchy of interior and exterior that preempts vision.

The Alteka Tower project begins simultaneously with an "el" shape drawn in both plan and section. Here, a change in the relationship of perspectival projection to three-dimensional space changes the relationship between project drawing and real space. In this sense, these drawings would have little relationship to the space that is being projected. For example, it is no longer possible to draw a line that stands for some scale relationship to another line in the space of the project, thus the drawn lines no longer have anything to do with reason, the connection of the mind to the eye. The drawn lines are folded with some ur-logic according to sections of a fold in René Thom's catastrophe theory. These folded plans and sections in turn create an object, which is cut into from the ground floor to the top. When the environment is inscribed or folded in such a way, the individual no longer remains the discursive function; the individual is no longer required to understand or interpret space. Questions such as what the space means are no longer relevant. It is not just that the environment is detached from vision, but that it also presents its own vision, a vision that looks back at the individual. The inscription is no longer concerned with aesthetics or with meaning but with some other order. It is only necessary to perceive the fact that this other order exists; this perception alone dislocates the knowing subject.

The fold presents the possibility of an alternative to the gridded space of the Cartesian order. The fold produces a dislocation of the dialectical distinction between figure and ground; in the process, it animates what Gilles Deleuze calls a smooth space. Smooth space presents the possibility of overcoming or exceeding the grid. The grid remains in place and the four walls will always exist, but they are in fact overtaken by the folding of space. Here there is no longer one plani-metric view which is then extruded to provide a sectional space. Instead, it is no longer possible to relate a vision of space in a two-dimensional drawing to the three-dimensional reality of a folded space. Drawing no longer has any scale value relationship to the three-dimensional environment. This dislocation of the two-dimensional drawing from the three-dimensional reality also begins to dislocate vision, inscribed by this ur-logic. There are no longer grid datum planes for the upright individual.

Alteka is not merely a surface architecture or a surface folding. Rather, the folds create an affective space, a dimension in the space that dislocates the discursive function of the human subject and thus vision, and at the same moment, creates a condition of time, of an event in which there is the possibility of the environment looking back at the subject, the possibility of the *gaze*.

The gaze according to Maurice Blanchot is that possibility of seing which remains covered up by vision. The gaze opens the possibility of seeing what Blanchot calls the light lying within darkness. It is not the light of the dialectic of light/dark, but it is the light of an otherness, which lies hidden within presence. It is the capacity to see this otherness which is repressed by vision. The looking back, the gaze, expos-es architecture to another light, one which could not have been seen before.

Architecture will continue to stand up, to deal with gravity, to have "four walls." But these four walls no longer need to be expressive of the mechanical paradigm. Rather, they could deal with the possibility of these other discourses, the other affective senses of sound, touch, and of that light lying within the darkness.

Visions unfolding: architecture in the age of electronic media is published by per-mission of *Domus* (no. 734, 1992, pp. 17-21).

Further reading

A book is based on other books and this one in particular tries to follow the numerous notes made during the course of a unique experience, starting in Eisenman's office in 1996 while working on the projects and continuing in 1997 while studying Eisenman's written works. Its contribution is not intended to be critical, but interpretative, and it refers to a method of work. The main studies used while writing this book are given in the following list; more specific references are included in the relative paragraphs.

Barry 96 – Donna Barry, *Eleven Authors in Search of a Building*, The Monacelli Press, New York 1996.

Ciorra 93 – Pippo Ciorra, *Peter Eisenman*, Electa, Milano 1993.

Davidson 96 – Cynthia C. Davidson, *Lotus*, no. 92.

Deleuze 72 – Gilles Deleuze, Félix Guattari, *Difference and Repetition*, Columbia University Press, New York 1994 (first published 1968).

Eis 1 – Peter Eisenman, *Palladio*, Lecture Series, unpublished manuscript, 1997.

Steiner 92 – George Steiner, *Real Presences*, University of Chicago Press, Chicago 1989.

Eis 2 – Peter Eisenman, *Forming the Critical*, Lecture Series, unpublished manuscript, 1996.

Eis 3 – Peter Eisenman, *Critical in Architecture*, Lecture Series, unpublished manuscript, 1995.

Eis 4 – Peter Eisenman, *Figuring the Ground*, Newsline, Columbia University Avery Hall NY, Nov./Dec. 1995, vol. 8, no. 2.

Eis 5 – Interview of the author, unpublished, 1997.

Eis 6 – Maurizio Bradaschia, interview in *Il Progetto*, no. 1, 1997.

Eis 7 – Peter Eisenman, *L'Architettura - cronache e storia*, no. 484.

Eis 8 – Alejandro Zaera Polo, interview in *El Croquis*, no. 83, 1997.

Eis 9 – Peter Eisenman, project meeting, author's diary, 1996.

Eis 10 – Peter Eisenman, project meeting, author's diary, 1996.

Eis 11 – Peter Eisenman, project meeting, author's diary, 1996.

Eis 12 – Peter Eisenman, project report, 1997.

Eis 13 - Peter Eisenman, "Vision Unfolding: Architecture in the Age of Electronic Media," in *Domus*, n. 734, January 1992.

Eis 14 - Interview in *Il progetto*, no. 1, 1997.

Krauss 96 – Yves-Alain Bois, Rosalind Krauss, *Formless: a User Guide*, catalogue, 1996.

Levy 97 – Pierre Levy, *Il virtuale*, Raffaello Cortina editore, Milano 1997.

Rockeborg 97 – *ANY Magazine*, n. 20, 1997.

Saggio 96 – Antonino Saggio, *Peter Eisenman. Trivellazioni nel futuro*, Testo & Immagine, Torino 1996.

Tafuri – Manfredo Tafuri, *La sfera e il labirinto*, Einaudi, Torino 1980.

CRITICAL ARCHITECTURE

When writing this chapter I used the concept of ongoing criticism elaborated by George Steiner in his book *Real Presences* cit. Listening to Eisenman's lessons, working in his office and attending the summer seminars made me realize the importance of a good system of analysis, a realization furthered by reading his Ph.D thesis *The Formal Basis for Architecture* which will soon be published. The transcriptions of the summer seminars, *Critical in Architecture* (1995), *Forming the Critical* (1996), *Palladio* (1997) can only be obtained from private sources, but they should be published by someone; I cannot honestly understand why Eisenman has not done so, or perhaps I do: the content of these lessons is translated into space in the realized and unrealized architectural projects.

Eisenman compared to Piranesi

Manfredo Tafuri's book *La sfera e il labirinto* and the essay on Piranesi, in the same book, by Sergej Ejzenštejn proved essential for writing this chapter. My reading of these works and a journey to Cincinnati with a few architect friends (thanks to Enzo, Riccardo and Alessandro) showed me every analogy and revealed Eisenman's inspiration. The Aronoff Center is itself a critical essay on Piranesi.

HOW EISENMAN WORKS

Much has been written on this architect. Before working in his office, I read the collection of essays entitled *La fine del Classico* edited by Renato Rizzi, CLUVA, Venezia 1987, and the monograph by Pippo Ciorra, *Peter Eisenman*, Electa, Milano 1993, whose introduction by Giorgio Ciucci *Ennesimeanamnesi* is one of the most interesting works on the American architect. In the same series as this book, I would also draw attention to *Peter Eisenman. Trivellazioni nel futuro* by Antonino Saggio, Testo & Immagine, Turin 1996, a genuine pocket guide to the American architect's activities. Renato Rizzi's *Mistico Nulla*, Motta, Milano 1996 is a very special and original work. For a clearer understanding of Eisenman's design process, see *Eleven Authors Searching for a Building*, The Monacelli Press, New York 1996, in which, having worked with Eisenman for several years, the eleven authors in the title describe and interpret one of the most complex buildings of contemporary architecture.

On other projects:
El Croquis, 1989, no. 41; 1997, n. 83, monographic issue on Peter Eisenman.
Eisenman Architects/ARX Geneve, *Building the Between*, Architekturgalerie, Munich 1997.
Industria delle costruzioni, 1998, no. 317, monographic issue on Peter Eisenman.
Greg Lynn, "Aronoff Center," in *Domus*, no. 788, December 1998.
Casabella, ott. 1996, no. 638; July/August 1998, no. 658.
"Formas de l'Informe," in *Arquitectura Viva*, Sept/Oct 1996, no. 50.
Luis Fernandez-Galiano, "Arte y arquitectura bajo el signo de Bataille," in *Arquitectura Viva*, Sept/Oct 1996, no. 50.
A+U Architecture and Urbanism, 1988 August, extra edition, *Peter Eisenman*.

INSTRUMENTS FOR ELECTRONIC ARCHITECTURE

When writing this chapter it was mandatory to refer to the works already published in this series, in the section The IT Revolution. By reading these, it is possible to start the journey into the electronic era:
L. Prestinenza Puglisi, *Hyperarchitecture. Spaces in the Electronic Age*, Birkhäuser Verlag, Basel 1999.
G. Schmitt, *Information Architecture. Architecture's new instruments, media and partners*, Birkhäuser Verlag, Basel 1999.

Special reference was also made to issue no. 3, Nov-Dec 1993 of *ANY Magazine*, edited by Cynthia Davidson, *Electrotecture: Architecture and the Electronic Future*. The following works were also consulted:
W. Mitchell, *City of Bits*, MIT Press, Boston 1996;
C. Boyer, *Cybercities*, Princeton Architectural Press, Princeton 1996;
N. Negroponte, *Being Digital*, Knopf, New York 1995;
M. Dely, *Escape Velocity Cyberculture at the End of the Century*, Grove Press, New York 1997;
Derrick de Kerckhove, *The Skin of Culture*, Somerville Press, Toronto 1995.

VIRTUAL HOUSE

The comments on the virtual were taken from Pierre Levy, *Becoming Virtual*, Plenum Trade, New York 1998, which compares the stances taken by Gilles Deleuze and Félix Guattari.
It is also important to refer to the work by Tomàs Maldonado, *Reale e virtuale*, Feltrinelli, Milano 1994. The description of the project is the outcome of the explanations and a careful reading of the article by Ingeborg Rocker "Virtual House" in *ANY Magazine*, no. 15, 1997.

THE NEW SYSTEM OF VISION

This chapter stems from a re-reading of Peter Eisenman's essay, published in the appendix: "Visions' Unfolding: Architecture in the Age of Electronic Media." Several years have passed since it was first published and Eisenman's architecture has undergone formal changes, undoubtedly backed by the new electronic instruments which have succeeded in not overwhelming his ideas and concepts. His design process has been refined using digital techniques and the endless search continues unchanged for a new type of vision.

On the folding theory: Eisenman Architects, *Unfolding Frankfurt*, Ernst & Sohn Berlin 1991.

Eisenman as a "machinic architect": A. Zaera Polo, "A conversation with Peter Eisenman: Eisenman's Machine of Infinite Resistance," in *El Croquis*, no. 83; and "Peter Eisenman, Processes of the Interstitial, Notes on Zaera Polo's Idea of the Machinic."

The term "machinic" is defined in an essay by Félix Guattari entitled *On Machines*. The definition is given in G. Deleuze, F. Guattari, *A Thousand Plateaus*, University of Minnesota Press, Minneapolis 1987.

I based my remarks on the Staten Island project on the correspondence with my brother, who worked on the project.

The Information Technology Revolution in Architecture is a new series reflecting on the effects the virtual dimension is having on architects and architecture in general. Each volume will examine a single topic, highlighting the essential aspects and exploring their relevance for the architects of today.

Other titles in this series:

Information Architecture
Basis and Future of CAAD
Gerhard Schmitt
96 pages, 70 color and 40 b/w illustrations
ISBN 3-7643-6092-5

HyperArchitecture
Spaces in the Electronic Age
Luigi Prestinenza Puglisi
96 pages, 60 color and 80 b/w illustrations
ISBN 3-7643-6093-3

Further titles will be published in the near future.

For our free catalog please contact:

Birkhäuser – Publishers for Architecture
P. O. Box 133, CH 4010 Basel, Switzerland
Tel. ++41-(0)61-205 07 07; Fax ++41-(0)61-205 07 92
e-mail: sales@birkhauser.ch
http://www.birkhauser.ch